THE AGRICULTURAL COMMUNITY IN SOUTH-WEST WALES AT THE TURN OF THE TWENTIETH CENTURY

THE AGRICULTURAL COMMUNITY IN SOUTH-WEST WALES
AT THE TURN OF THE TWENTIETH CENTURY

BY

DAVID JENKINS

CARDIFF
UNIVERSITY OF WALES PRESS
1971

PRINTED IN GREAT BRITAIN

I'M RHIENI
A'U CENHEDLAETH

PREFACE

THIS book has twin origins. It is based on material which I prepared for presentation to classes conducted in West Wales by the Department of Extra Mural Studies of the University College of Wales, Aberystwyth. I was then seeking for relatively familiar examples whereby I might illustrate certain of the ideas which are the stock-in-trade of anthropologists and which might be more readily intelligible than less familiar material.

But what got the book under way was the conversation of the people of the area, their ready help and their evident interest in the society that is discussed in these pages. My first thanks are to them. If any one of them is to be mentioned then it must be Dan Pwllgwair, who did not live to see this book in print.

I have been greatly aided by all those who have read the manuscript for me, by Professor Alun Davies of the University College of Swansea; by Professor E. L. Peters and by Mr. B. L. Sansom of the University of Manchester, and by Dr. Richard Phillips of Llangwyryfon. They went to great pains to help me and I am glad of the opportunity of acknowledging my obligations to them. My thanks too to Professor E. G. Bowen of the Department of Geography of the University College of Wales, Aberystwyth, and to Mr. Morlais Hughes, for help with the preparation of the maps and diagrams included herein. I also acknowledge gratefully the aid given to publishing this work by a generous subvention from the Trustees of the Catherine and Lady Grace James Foundation.

And my particular thanks to my wife without whose help this book would not have been completed at all.

DAVID JENKINS

CONTENTS

PLATES

(following page 280)

FIGURES IN TEXT

TABLES

I

INTRODUCTION

THERE is no need to labour the change that has occurred in both urban and rural areas during the lifetime of people now in old age. This study is concerned with a rural area where agriculture was and is the most important occupation. During the youth of those who are now old all who lived in this area were, for reasons to be discussed, familiar with farm work and practice. This is no longer so. Many of those who live in the countryside excepting farmers now know little more of farm life than do town dwellers. As urbanization proceeds yet fewer will be familiar with life and work on the land.

For these reasons I shall be concerned to record and to describe the rural society that obtained in this particular area. I shall not beg the questions involved in labelling it a 'peasant society', but I hope to show how the form of the society was related to the needs of working the land. It remains possible to talk with people who remember rural life as it was before the introduction of many of the machines which have served to change much of that life, the corn binder, the mowing machine, hay-harvesting equipment, the potato digger, the tractor. This is not to say that it is possible to return to a period before any change began, but it is possible to return to a time when the relationship between the society and the land was substantially different from that which now obtains.

The field work was undertaken during the years 1958–61. My oldest informant remembered the coming of the first horse-drawn threshing machine in or about 1875; other informants started work on farms during the 1880s. The informants included farmers, farmers' wives, labourers and servants, maids, craftsmen, teachers, ministers of religion, civil servants, and other professional people. I have been concerned to learn from them matters of fact, for example at what age did they first use the scythe at the corn harvest? On what occasions did farms co-operate? Which farms formed co-operative groups? It has been possible to cross-check extensively and I am of the opinion that informants can provide accurate information, particularly about farming activities, from the time that they were fifteen years of age or thereabouts. It was at this age that they would have left school, started full-time work on the land, first taken their places in harvest work-groups as scythemen, an occasion none has forgotten, become full members of religious denominations, and entered adult society. This means that the mid 1890s are the first years for which there is extensive information, and I shall be concerned to describe the society as it then was and to indicate the changes that were occurring at that time.

I note certain other considerations that have helped this reconstruction. The study is concerned with a literate society for which there is a wealth of documentary evidence in the form of official reports, minute books and statistics, extensive press coverage, and private documents. It was, too, a society with a long popular literary tradition, and the sources are augmented by numerous essays submitted to local competitive meetings (*eisteddfodau*), diaries, letters, and the like. (See note on p. 36.)

One of the distinctive features of the community was that it was 'parochial'. The word is not used in a pejorative sense. 'Parochialism' does not consist of nothing but a lack of interest in general affairs, indeed it is not necessarily inconsistent with such an interest. Its positive content is an interest in one's own society, in one's neighbours and their affairs, their kinship links, family histories, marriages, fortunes, their movements from one farm to another. Many local practices have reinforced people's memories. It has been common practice to refer to a man by his Christian name and by the name of his farm, as, for example, *Wil Gwndwn*, i.e. William of *Gwndwn* Farm, so that a man who has moved several times to a different farm may be variously known by his Christian name and the name of one of several farms, among them that which he may have occupied sixty years earlier. Should someone leave the locality to work elsewhere and return to the district of his birth many years later to spend his retirement, he may be known by the name of the house in which he lived before migrating. Hence one is aided in the attempt to discover the details upon which to base a reconstruction.

It may be noted that remembering details of the past commonly consists not in recalling events known or experienced only once, but remembering facts which have been periodically recalled. Thus on inquiring about farm practices which have disappeared one is inquiring about practices which were repeated consistently year after year.

Moreover, some practices that have generally disappeared still continue on some very small holdings or in remote places of difficult terrain, or under unusual circumstances. In 1953 I counted

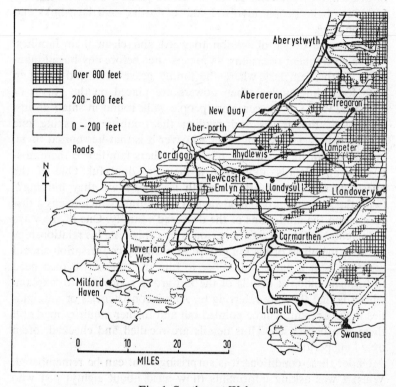

Fig. 1. South-west Wales

thirty-five people engaged in the hay harvest of an 8-acre holding. They worked from midday till seven o'clock in the evening with none of the equipment that has long been common on larger holdings. (I saw the same holding's hay harvested in 1960, when a baler and lorry were used: in two hours six men baled all the crop and carried it away.) During the summer of 1959 I watched men cutting corn with a scythe and binding it by hand in a North Cardiganshire valley within twelve miles of Aberystwyth. As late as 1951 the reaping hook was used in mid Cardiganshire to save the crop during a difficult harvest.

Inquiries about kinship links are also inquiries about facts that will have been recalled on numerous occasions as when marriage takes place. People then ask questions about the kin relationships of bride and groom, and this turns into a general discussion of the fortunes of the people involved, and of further marriage links and relationships.

People are under obligation to weed and clean their families' graves on frequent occasions, as for instance before any burial takes place in the cemetery where the family graves are, and again in time for Palm Sunday when flowers are placed on the graves of relatives. While in the cemetery people walk round in small groups looking at the graves and reading the tombstones, asking one another whether they remember whoever it is that lies buried there, inquiring how he or she was related to others familiar to them, discussing any details about which they are doubtful. One of the questions soon asked after a death is 'Where are they burying?' (with the personal pronoun always in the plural), which opens a discussion on the affiliations of the deceased's family. One of the ways of 'placing' or 'identifying' a person is by explaining his relationships to other members of his kin group. In ordinary conversation should someone hear a name unfamiliar to him the others will then need to furnish sufficient details of the kin group concerned to 'explain' who it is that they are referring to. Any inconsistencies or ambiguities in the account will be pointed out and further inquiries made till there is a consensus. Thus details are recalled and checked, often over long periods.

Under these conditions it is surprising what can be remembered. When I was asking informants (now aged about eighty) just who had occupied particular farms in the past, it was necessary to inquire carefully what period they were referring to, for on several occasions I found that they started with people who had occupied those farms during the early nineteenth century. On another occasion a newspaper containing an account of the opening of a nonconformist chapel in 1832 was, in my presence, shown to a man who was in his sixties. The account named twenty-one people; the informant identified nineteen of them as 'so-and-so's' great grandfather or whatever else the connection was, and disclaimed knowledge of the other two. It may be stressed that it is possible to check a great deal of information collected in this way, and the following example is provided. I asked one informant aged eighty, who had spent his whole life in one locality, what people had occupied each of the 108 agricultural holdings in his parish in or about the year 1900. His information was checked against the parish electoral list for that year, and found correct in respect of 106 of the holdings. Finally I make it clear that I know that not all people can remember thus; what is important is that some can. In this society what is most accurately remembered is what has been periodically recalled and this is usually material concerning people; memory is much less reliable when it concerns events that neither recurred nor were consistently recalled.

It has been stated above that the concern of this study is to record and describe the society that obtained in a particular rural area. But this presupposes selection and analysis of material, otherwise nothing would be provided but a mass of unrelated items of information, a sort of social antiquarianism. Instead it is hoped to study the structure of the society, and the changes in that structure during the lifetime of those who are now aged. In this society people stood related to one another in various and characteristic ways. As members of the staffs of individual farms they stood related as farmers on the one hand and farm servants and labourers on the other. People stood related too as farmers and cottagers in their capacities as members of the work groups which were connected with every farm. They again stood related to one another as members of places of worship, whether as members of the one place of worship or as members of different but not unconnected places of worship. The structure of the society is here conceived of in terms of the institutionalized and interconnected roles performed by individuals who stood in such various relationships to one another in virtue of their membership of such groups as have been noted above, and whose significance is to be shown below. Changes in the structure of the society are considered in terms of changes in these roles and in terms of the associated changes which occurred in the groups as the people who comprised them modified the ways in which they performed their roles, ceased performing some roles, and evolved new ones in the course of time.

This does not include the whole wealth and range of variation in individual personal relationships, which in any case are no longer observable. This in turns puts a practical and logical limitation on the work which thereby becomes that of making a model of the society that has existed in the area with which the study is concerned. It has been indicated that there are adequate sources, oral and documentary, for this task, the limitations that such sources impose being the necessary condition of doing the work at all.

South Cardiganshire is a land of dispersed farms and small villages, with a number of market towns in the Teifi Valley. Within this area it is not possible to draw geographical boundaries around any community selected for study, as might be possible if settlements were nucleated in villages with open country between them. A holding of 150 acres was, and is, considered a large farm, hence farm-houses stand relatively near to one another and from any one dwelling-house there are various others in sight. Everywhere these farms formed co-operative groups, in such a way that groups were not exclusive but overlapped one another. Each farm was the centre of its own particular group, farms A and B for instance being members of one group, B and C (but not A) members of another. Hence there was a network of co-operating farms. Groups of cottages were connected with each farm in this organization of overlapping farm groups, and membership of these farm-cottage groups might overlap in the same way that groups of farms overlapped. Hence no boundaries can be drawn that would isolate one community from all others.

When schools were built to implement the Education Act of 1870, they were frequently built not in the villages but between them, in order to minimize the number of buildings required, though this would scarcely have been done had there been any strong feeling that the villages *ought* to be the foci of activities. Churches and chapels may stand in the villages or away from them, and their members neither were, nor are, necessarily the people that lived or live nearest to them. Farms, churches, chapels, schools are all scattered, all of them foci for distinct activities, and all drawing on people living alongside others who have different centres for any, or all, such activities. It is a common experience for people living near to one place of worship to meet others walking to that place of worship while they themselves are on their way to another place of worship. Prior to the implementation of the 1944 Education Act it was common for children on their way to one school to cross the path of those on their way to another school, and common for their parents to cross each others' paths on such occasions as school concerts. At harvest, and other times, cottagers on their way to one farm met others on their way to other farms. That is to say, here are no cell-like neighbourhoods each with its own nucleus, but groups that overlap territorially and functionally.

At the same time people recognized themselves as members of one neighbourhood rather than another, such neighbourhoods deriving their names from a village, a church, a prominent physical feature,

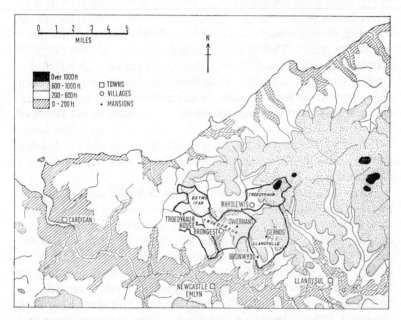

Fig. 2. South Cardiganshire

as the case may be. But now, as in the past, a neighbourhood does not have one centre for all of its activities. Rather, there are different centres for different activities, usually there are several centres for each activity, and they are not necessarily the centres for those who live in that particular geographical neighbourhood. In other words, while people think of themselves as members of a particular neighbourhood they also think of themselves as members of kindreds, churches, chapels, and co-operative groups which need have no particular connection with their neighbourhood. Membership of a neighbourhood and membership of groups which have no exclusive territorial basis are not mutually incompatible.

Just as no geographical boundaries can be drawn between communities within South Cardiganshire, neither can a distinct boundary be drawn between South Cardiganshire and adjacent areas. Many activities, such as those of local and central government, involve the people of South Cardiganshire (as of all other areas) in a very much wider sphere of activities. These, far from being unimportant because they are not distinctive of a particular area, are in fact among the conditions that allow any area to be such as it is. Nevertheless within South Cardiganshire there is sufficient in common, though it may also be common to other areas, to provide a subject for study. That this is so is indicated by the movement of people within the area. Certainly, during the period with which this study is concerned (the lifetime of those now in old age) it has been very common for people to move from one farm to another. Farmers' children may need to move considerable distances to get farms of their own for the first time; later they may move to better farms when the opportunity arises. As a result, there are links of kinship and friendship, there are standard farm practices and their associated social institutions, in this area.

Hence the study relates to South Cardiganshire in general, but one parish, Troedyraur, has been chosen for the most detailed investigation. Figures 3 and 4 show the places to which members of two chapels (Hawen and Twrgwyn) in the village of Rhydlewis in the parish of Troedyraur, moved during the period 1899 to 1926. This can be done accurately for letters transferring their membership from these chapels to others were given when members moved. The details were recorded in the chapels' records, and these have been consulted. What is immediately relevant is that the map shows that population movement (though this map shows but part of it) links this parish with the rest of the area that is under consideration.

The practical reasons for limiting detailed investigations to one parish will be obvious but there are other considerations to bear in mind. People living beyond the confines of the one parish and beyond the general area of South Cardiganshire share many of the features of the social structure which is to be found in that area. Certain of the relationships that obtain between people living in South Cardiganshire, such as those involved in the structure of farm families, are found widely in Wales; others, such as the relationship between farmers and groups of cottagers, do not obtain over such a wide area. Thus the study is less concerned with the exact

boundaries of the areas of either the detailed or the more general
investigation, as with the structure of a society that does not consist
of discrete communities.

In this study use has been made of idioms and sayings that are in

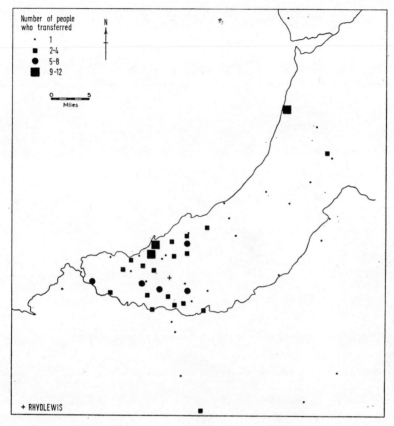

Fig. 3. Location of chapels in south-west Wales to which members of chapels in the
village of Rhydlewis transferred their membership, 1899–1926

frequent use. The fact that they are commonly used implies that they
have referents that are commonly acknowledged, and as they are
indigenous they provide a guide to the categories according to
which people conceptualize their own affairs. This means that the
society is described in the first place, according to the terms in
which its members see it rather than according to categories devised

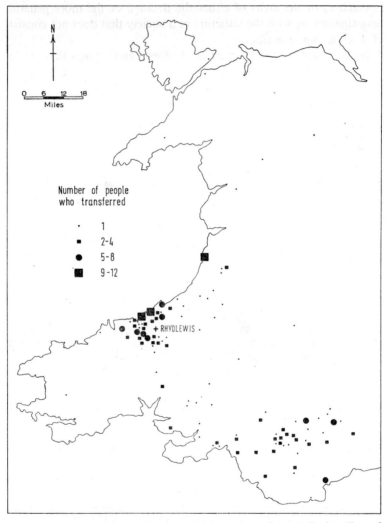

Fig. 4. Location of chapels in Wales to which members of chapels in the village of
Rhydlewis transferred their membership, 1899–1926

by the investigator. It is clear that such terms do not provide a complete and ready-made description of the society, for obsolescent
terms sometimes remain in use when their referents are no more
than memories, while, on other occasions, specific and distinctive
terms may be lacking. Further to depend entirely on local sayings

would limit a description of the society to those aspects which its members do conceptualize. Nevertheless, local idioms can provide a valuable aid to social analysis.

There is a wealth of such terms. A fat boy is said to be 'like a calf sucking two cows', while one needs to 'raise the feeding rack' of a corpulent man. (The Welsh originals are given in the footnotes). Someone who will not work except in the limelight 'wants to be a lead horse'; one who speaks fluently is 'like a sea pool'. If he speaks sonorously while his words lack substance he is 'like a bee in an urn', and should his claims to leadership be pretentious he is a 'wooden deacon'. If an insignificant person he is 'like an acorn in a sow's belly'. In the case of sudden death a man departs 'between hand and sleeve', while the expression for a gentleman-only funeral is a 'men and servants funeral'. Thin slices of bread and butter are 'preacher's bread and butter', while setting up house by a newly married couple is spoken of as 'starting their world'.[1]

Many of these expressions are easily collected and can be classified in various ways, for example according to the origin and development of each phrase. Some derived from farm practice, others from the sea, and yet others from chapel associations. This is the method employed in such a book as Logan Pearsall Smith's *Words and Idioms*.[2] A linguist will classify them differently into adjectival phrases, noun phrases, etc., according to their syntactical function.

They may be classified too, according to their common function in everyday conversation, thus:

1. Ejaculations, which are frequently untranslatable.
2. Proverbs, e.g. 'The people of the little houses kill a pig but once a year'.
   'Black Saint Cynon's fair, the cows to the byre.'
3. Descriptive sayings, such as many of those noted above.
4. Denotative sayings: that is, sayings that do not simply describe but denote types or classes of things, actions, and persons. The phrase 'two-ended house', for instance, not only describes particular houses but denotes a house type; 'red land' denotes a certain type of land use, land under corn.

Abbreviations: H.M.S.O.   Her Majesty's Stationery Office
                N.L.W.   National Library of Wales

[1] *fel llo yn sugno dwy fuwch, codi'r rhastal, eisiau bod yn geffyl blaen, fel pwll y môr, fel cachgi mewn sten, daeacon pren, fel mesen ym mola hwch, rhwng llaw a llawes, angladd gwŷr a gweision, bara menyn pregethwr, dechrau'u byd.*

[2] L. Pearsall Smith, *Words and Idioms, Studies in the English Language* (London, 1925).

What is of immediate consequence is that some of these terms refer to categories of the society, terms which parallel what Nadel referred to as 'the diacritical use of names': 'Social status and the membership of groups are rendered visible by the diacritical use of names . . . whose every application will indicate whether a particular person "belongs" or does not "belong"; to trace the repetitive or varied use of these names is thus to trace the social framework.'[3]

During the period that the study is concerned with people spoke of *y gwŷr byddigion*, that is the gentry, *y gwŷr mowr*, literally, 'the great people', '*y* (the) blue bloods', these terms being synonyms for one category of the society. Another category is indicated by the term 'people of the little houses' (*pobol tai bach*), a 'little house' signifying a house with or without garden but with no land attached, unless sometimes with one or two fields where a cow or two were kept to provide 'milk and butter sustenance'. This is the social category referred to in the above mentioned proverb '*The people of the little houses* kill a pig but once a year', which refers to the limited resources of these people. A standard work on North Pembrokeshire dialect explains the Welsh word *glosced* thus—'Large tracts of furze are set on fire in March, and the charred stumps are used for firewood by "*y bobol bach*' (the poor people) in winter. Farmers readily allow the poor to go on their land to gather *glosced*'—thus distinguishing between '*y bobl bach*' (literally, 'the little people') and the farmers.[4] Well within living memory school children in some Teifi Valley schools divided 'naturally' into *plant ffermwyr* (farmers' children) and *plant tai bach* (children of the little houses) for team games in the playground.

That '*pobol tai bach*' was a widely recognized category is shown by the widespread distribution of this or a corresponding term. In certain coastal areas of Cardiganshire the term in use was 'people of the clay houses' (*pobol tai clai*), in Carmarthenshire 'men of the little houses' (*gwŷr tai bach*), while in parts of Merioneth and Montgomeryshire *pentŷ* (which is untranslatable) corresponds to 'little house' and *pentŷaeth* (again untranslatable) to 'people of the little house'. In Caernarvonshire village houses were referred to as 'little houses' (*tai bach*), and in some areas a 'bare house' (*tŷ moel*) corresponded to 'little house'. As these terms are cumbersome

[3] S. F. Nadel, *The Foundations of Social Anthropology* (London, 1953), pp. 45–6.
[4] W. M. Morris, *A Glossary of the Dimetian Dialect* (Tonypandy, 1910), p. 145.

having presented them it will be possible henceforth to refer to the 'people of the little houses' as 'cottagers' wherever that is practicable.

Gentry and cottagers constituted two categories of the society, 'farmers' constituted a third. They were referred to as such rather than by a particular local expression. These three categories are well illustrated by this quotation from a cottager's diary which refers to the coming of age of the heir to the Alltyrodyn estate in 1871, 'Today is a merry festival in this locality for all of Alltyrodin's tenants; the heir to the estate came of age, Tomos and all the other tenants, farmers and *bobl fach* (lit., little people) went to Newinn to dinner . . .'[5] As can be seen, the categories between which the diarist differentiated were the gentry, the farmers, and the cottagers. These categories constituted the basic stratification of the society. The gentry were soon to disappear, and the relationships between the farmers and the cottagers were soon to change.

Physiography

The Teifi Valley runs roughly parallel to the coast for much of its length, before the river turns west towards the sea. Between the river and the sea stands a ridge, wider and higher to the north-east, narrower and lower to the south-west. Its crest is rounded but un-even, rising to moorland hills of over a thousand feet above sea-level in the north-east while the hill tops are progressively lower towards the south-west where the ridge finally declines into the vale of Teifi. The higher hills are exposed to the sea winds, their soils are acid and peaty, and tree growth is stunted. Where the hill tops are lower the moorland character is less obvious or absent, the soils less acid and of greater depth. But everywhere, at elevations greater than 500 feet, the natural vegetation tends to heath. The country rock is generally shale or mudstone, limestone being quite absent so that the soils are deficient in lime. Glacial clays are irregu-larly distributed throughout the area, making valley bottoms and level moorland expanses boggy, but providing moderately fertile loams on the better drained slopes. In contrast soils are much lighter and thinner and rather liable to drought where there are no glacial clays.

From the crest of the central ridge short rivers run west to the

---

[5] Unpublished manuscript in private possession.

sea through steep-sided valleys, the valley bottoms being clayey, ill-drained, and rushy. Inland from the central ridge of South Cardiganshire run the tributaries of the Teifi. Their valleys are generally narrow, their sides occasionally too steep for horses to plough and the bottoms reed-strewn and boggy. Thus the better farm land is on the extensive intermediate slopes. As the whole area, and particularly the seaward slope, is open to Atlantic winds and to rain, slope aspect and exposure are important differentiae, especially as the higher and more exposed ground has the more acid soil. In combination a loam soil, a measure of shelter from wind and rain, a southerly aspect, and good drainage provide the best farming conditions, deteriorating on the one hand to upland heath, and on the other to boggy bottom land.

These surface conditions in conjunction with a mild and moist climate are more favourable to grass than to arable farming. Some 40 inches of rain fall on the lower ground per annum, and over 50 inches on the higher ground. But in the earlier part of the period under consideration, when wheat was grown 'for the use of the house', i.e. as bread corn, the annual variation in rainfall incidence during the growing season and at harvest time was itself a major difficulty.

Rainfall records which were kept at a recording station in the parish of Troedyraur (in South Cardiganshire) are available for the years 1889 to 1895 (inclusive) and the relevant data are presented in the accompanying table.

The first cereal to be sown during the spring was oats which was sown during February if that was possible. Ideally all cereals should have been sown by the end of April. The months of March, April, and May thus constituted the early growing season. July was the time when the hay harvest began, while August and September were the months of the corn harvest. It will be seen that rainfall is liable to vary from close to double the mean rainfall to less than a quarter of the mean rainfall in the course of the growing season, from half the mean rainfall to half as much again during the hay harvest period, and from a quarter of the mean rainfall to one and two-thirds of the mean rainfall during the corn harvest months. Six times as much rain fell during the early growing season of 1895 as had fallen during the same period in 1893. During the main corn harvest month of 1891 more than six times as much rain fell as was to fall during the corresponding month of 1895. This variability

TABLE 1. *Monthly rainfall at Troedyraur Rectory, 1889–1895 inclusive*[6]

	Mean monthly rainfall, 1889–95 (in inches)						
	Mar.	Apr.	May	June	July	Aug.	Sept.
	1·87	1·84	2·43	1·93	3·10	3·4	2·3
	Maximum monthly rainfall, 1889–95						
	Mar.	Apr.	May	June	July	Aug.	Sept.
Year	1895	1895	1889	1891	1895	1891	1891
Rainfall (in inches)	3·48	3·47	3·60	3·51	4·91	4·53	3·85
Percentage of mean	186	188	148	172	158	133	167
	Minimum monthly rainfall, 1889–95						
	Mar.	Apr.	May	June	July	Aug.	Sept.
Year	1893	1893	1895	1895	1892	1895	1895
Rainfall (in inches)	0·76	0·40	0·50	0·53	1·60	2·37	0·59
Percentage of mean	41	23	21	28	52	70	26

clearly makes for unpredictable conditions during the growing season and at harvest times, and this was even more important before silage-making became a common practice, and before hay-sheds and harvesting machinery were introduced, than it is today. It is therefore relevant to a discussion of social structure because the hay and corn harvests were critical when people relied far more heavily than they now do on local production for food for themselves and feeding stuffs for their stock.

With a minimum mean monthly temperature of 41·6 °F. (February) and a mean maximum of 60 °F. (August) there are always wild plants in bloom where there is shelter. Ash, sycamore, and oak grow strongly, with hazel, willow, and alder in the moister areas. Water meadows are rare for the valley bottoms are confined and prone to rushes, but soil and climate are favourable to grass growth on the better land. Hence dairying and stock-rearing characterize the area's agriculture and as the country is not particularly rich, farming has depended on rearing store cattle to be sold for fattening

[6] H.M.S.O., *British Rainfall*, 1889–1895 inclusive (London 1890–6): Monthly Rainfall Tables.

elsewhere rather than on fattening beasts for slaughter. This is consistent too with the area's relative remoteness from urban and industrial markets, a factor of particular importance before the development of motor transport, and, with it, the sale of bulk milk.

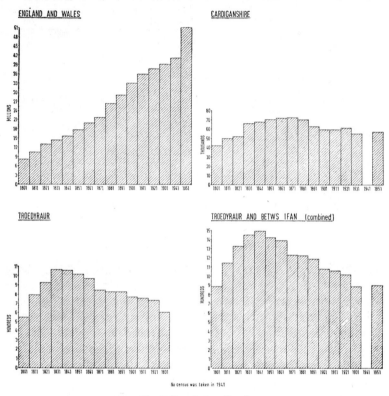

Fig. 5. Population Graphs

For more detailed study the parish of Troedyraur has been chosen. It includes a small valley basin where the moorlands to the north-east give way to better conditions towards the south and west. Here, at the confluence of the rivers Ceri and Dulais the valley floor is wider than is usual in streams flowing to the Teifi. The bottom land, as is common in these valleys, is rather ill drained as those farm names containing the element *gwern* (alder), Gwernant, Gwernddafydd, Bwlchywern, Pen-wern, suggest. At either end of the valley floor stand two villages, Rhydlewis and Brongest, some 300 feet above sea level and a mile and a half apart. Steep slopes rise to moorland ridges of 900 feet to 1,000 feet above sea level within the mile, on the south-east and east sides. To the north and west the slopes are gentler and well drained, rising to rounded hills which, at some 500 feet above sea level, are quite free of moorland. The larger village, Rhydlewis, is six miles or so from the market town of Newcastle Emlyn, and a comparable distance from Llandysul which is the area's other market town. These towns are too small to constitute markets in themselves, and the area is remote from large urban and industrial centres.

Figure 5 shows the population that has inhabited this parish. As the common boundary between this and the adjoining parish was changed between the census of 1921 and that of 1931 without affecting their other boundaries, it is their combined population (Fig. 5) that shows accurately how the population has changed over the years. The two parishes are alike in their physical geography, in their agriculture, and in their social composition. As is common in purely agricultural parishes in south-west Wales the population declined consistently from 1841 until the Second World War. But even with a slight recovery by 1951, the population was then little more than it was in 1801 when the first census was taken. This study aims to investigate how the social life of people who were principally dependent on agriculture was related to the needs of working the land that has been described.

The Gentry

In the different regions and different countries of Europe, many systems of landownership have existed, ranging from peasant proprietorship of small holdings to the landlord-tenant system of estates which had become characteristic of Great Britain well before the end of the nineteenth century. As then found, the system in Britain was not of great age. It took its modern form during the period of enclosures, furthered by the effects of the Napoleonic Wars. It was then that the yeomen were bought out along with smaller proprietors, and a much greater part of Britain passed into estate ownership than was previously the case. The system of gentry-owned estates was to be found in south-west Wales as elsewhere, and here it is proposed to note certain characteristics of the landlord-tenant system as it obtained at the end of the nineteenth century in the region that particularly concerns this study.

At an earlier period, during the seventeenth century, one characteristic of Wales was that in addition to the more substantial gentry families, there was a large number of minor gentry possessed of small estates:

the term 'gentry' as understood in Wales was a highly elastic and comprehensive one, embracing all who could put up any sort of colourable claim to descent from the old princes and *uchelwyr* (free born) and even some who had lifted themselves out of the servile status of their ancestors by judicious purchase, lease, or marriage during the long years when tribalism was in decay and the land market kept fluid by the frequency of civil commotion.[7]

Such minor gentry lived in what were by the standards of the time small mansions, but which by the standards of the late nineteenth century were substantial farm-houses though some of them still retained, and do today retain, the name *plas* (mansion).

During the eighteenth and much of the nineteenth centuries, in times of agricultural prosperity, many of the mansions were rebuilt, as Bronwydd and Gernos in the locality of Troedyraur. (Gernos was rebuilt during the Napoleonic Wars, and again during the early 1870s, while the complete rebuilding of Bronwydd began in 1854 and was completed a few years later.) In other cases new mansions were built, the erstwhile gentry homes being let to tenant farmers. Thus when the Walters family of Bedd Geraint built the small mansion of Glandmedeni their previous home became that of a tenant farmer who was not accounted a member of the gentry.

Many estates were acquired or enlarged by piecemeal acquisition of individual farms and of small estates, by purchase or marriage. Thus late in the seventeenth century the estate of Pantyrodyn (consisting of four farms each with its various parcels, totalling 644 acres) in the parish of Troedyraur passed to the Llannerch Aeron estate on the marriage of Margaret, daughter of Nicholas Lewis, Esq., of Pantyrodyn to John Parry of Llannerch Aeron. The one-time mansion of Pantyrodyn was then let to a tenant, and by the late nineteenth century not thought of as anything other than a substantial farm-house.

When Thomas Colby, a younger son of the Colbys of Ffynone and Rhosygilwen (in North Pembrokeshire) came of age, he inherited and eventually came to reside at Pantyderi, a minor gentry residence that had not been extensively rebuilt. H. M. Vaughan, Esq., one of

[7] A. H. Dodd, *Studies in Stuart Wales* (Cardiff, 1952), p. 1. Quoted by permission of the author and the University of Wales Press.

the 'Tivy-side' gentry, described Thomas Colby and Pantyderi during the early years of the twentieth century:

the squire, Mr. Thomas Colby, one of the Colby family of Ffynone and Rhosygilwen (and the son of old General Colby, an austere, distinguished soldier who refused a baronetcy), presided over a household that had many points in common with the establishment started by Count Tolstoy in Russia. All was simple to the verge of discomfort. Pantyderi was, in short, a mansion run on the lines of a farm-house, where all the men and maids dined with the family and all partook of the same coarse but abundant fare . . . Mr. Colby, generally called 'Twm Colby', was of course, a character, . . . He was indeed a quaint denizen of the countryside, wearing ordinary labourer's dress at home and appearing in black broad-cloth on other occasions, so that with his shiny black clothes and his 'Newgate frill' he might easily be mistaken for some old-fashioned dissenting minister.[8]

Thus by the end of the nineteenth century some minor gentry families, and branches of better-known families, lived in houses that were little, if at all, different from some of those occupied by substantial tenants. But this did not cloud the distinction between gentry and others, for the most minor or eccentric of gentry was the blood relative of other members of the gentry, in which social stratum a very high value was placed on pedigree and descent. Marriage between the gentry and the yeomanry was virtually unknown in the late nineteenth and early twentieth centuries, though it did occur on occasion in South Cardiganshire during the early years of the nineteenth century.

Another result of the formation or enlargement of estates by piecemeal acquisition, by adding single farms and such small estates as Pantyrodyn to nuclei which were themselves in many cases small, was that estates were constituted not of continuous blocks of land but consisted of single properties and small blocks of property scattered among similar properties held by other landowners. Hence in the parish of Troedyraur, late in the nineteenth century and before estates were sold on any considerable scale, twelve squires shared the ownership of the 2,400 acres in the parish which were owned by the gentry. (Other people owned the remaining 1,664 acres within the parish, some of them being occupiers as well while the others let their farms to tenants. Approximately half of the acreage owned by people other than the gentry consisted of the moorland which is to be found on the higher ground

[8] H. M. Vaughan, *The South Wales Squires* (London, 1926), p. 91.

in the north and east of the parish.) The twelve properties owned
by the gentry were not twelve compact blocks, but included scattered
individual farms. Other South Cardiganshire parishes showed a
similar state of affairs; seventeen squires owned 2,718 acres in the
parish of Blaen-porth, while nine squires owned 2,117 acres in the
parish of Aber-porth. In consequence no one squire dominated a
district, rather there were many squires constituting the 'county'
and at the same time in competition with one another for public
office and local influence.

There were two mansions in the parish of Troedyraur, namely
Troedyraur House, and Gwernant, whose name was used in its
translated form, Alderbrook Hall ('Alderbrook' being a literal
rendering of *Gwernant*), by the gentry who resided there during
the nineteenth century. But as neither of these houses was continuously occupied by its owner the more influential houses were
those of Gernos and Bronwydd which stood in the adjoining parish
of Llangynllo.

Gwernant was the seat of the Lloyd Williams family, John
Lloyd Williams, Esq. (who was High Sheriff of Cardiganshire in
1805) having built a new mansion on the upper slopes overlooking
the vale of Troedyraur to replace the old house that stood near the
foot of the slope. The estate in south-west Wales consisted of 646
acres in the vale of Troedyraur and other scattered properties totalling some 1,200 acres in all in 1893, with an annual gross rental of
£1,147. 16s. 0d. But from 1860 to 1900 very rarely did a member
of the family live in the mansion. Instead it was occupied by a series
of tenants who held sufficient land to keep riding horses and coach
horses, but without a home farm or other estate in the locality. J. C.
Yorke, Esq., a retired army officer from Middlesex, and his wife
(who was from Somerset) occupied the mansion during the 1860s,
though Caulfield Tynte Lloyd Williams, Esq., the owner, returned
briefly at the end of the decade. He died without issue and a life
interest in the estate passed to his widow who survived until 1898,
but removed from Gwernant which was again let to tenants. John
Boultbee, Esq., J.P., was in residence by 1871, his wife being a
daughter of William Lewes, Esq., of Llysnewydd, which stands in
the vale of Teifi. A retired army officer, James Graham Kell, Esq.,
followed him and resided there till the new owner came to Gwernant
soon after coming into the inheritance on the death of Mrs. Lloyd
Williams.

He was Gilbert Lloyd Williams, Esq., whose father and Caulfield Tynte Lloyd Williams were cousins. They were the sons of two
brothers. Gilbert Lloyd Williams was an Indian tea planter when
he came into the estate. On arrival at Gwernant he, a bachelor,
installed himself in one of the lodges and sold the mansion along
with 49 acres of land (17 acres of them being woodland) to one
Forde Hughes, Esq.

Forde Hughes was a most extraordinary person. He was the son
of Thomas Davies, Esq., of Nantgwylan, a house that Thomas
Davies himself had built and which stands in the same locality. Its
plan was characteristic of many substantial South Cardiganshire
farm-houses and it stood a mile distant from Gwernant on the one
side and Gernos on the other. Thomas Davies married a daughter
of Col. Owen Lloyd, a member of a younger branch of the Lloyds
of Bronwydd, whose mansion stands some two miles from Nantgwylan. There were two children of the marriage, a son who took
the name of Forde Hughes, and a daughter. Forde Hughes spent his
early life in London and Paris and although he succeeded to the
Nantgwylan estate in 1866 and was 'a wealthy landowner with one
or two places of his own', he finally settled in 'a mean house' in
Union Street, Carmarthen. He 'had a perfect mania for both buying
and renting vacant country houses. For years he was a perfect God-
send to the impoverished gentry of the Tivyside, for he would apply
for the refusal of any house that happened to be for sale or to let'.[9]
'Forde Hughes was never seen by daylight, but occasionally he
would hire a fly at nightfall and drive long distances through the
darkness to visit one or other of his various country seats, arriving
at some unearthly hour and only staying for a few minutes, in order
to reach his Carmarthen house again before daylight.'[10]

While the owner of a small mansion which stands in the vale of
Teifi (Aberceri, which he sold in 1873) Forde Hughes appears to
have installed a light railway in the very restricted grounds. While
the owner of Gwernant he imported an organ from Paris. This was
transported from the nearest railway station in wagons part by
part and assembled in the mansion which Forde Hughes hardly
occupied. He lived in Carmarthen in squalor; the sanitary conditions
of his house were noted as 'indescribable, especially after his water-
supply was cut off by the authorities'. The door was forced after

[9] H. M. Vaughan, *The South Wales Squires* (London, 1926), pp. 91–2.
[10] Ibid., p. 92.

food left for him was found untouched, and he was taken to the
Poor House where he died shortly afterwards during the spring
of 1914 at the age of sixty. He left £74,000.[11]

In the meantime Lloyd Williams, Esq., who had sold Gwernant
to Forde Hughes, continued to occupy one of the mansion's lodges.
In 1913 he sold the greater part of his estate but retained the estate's
home farm. Shortly afterwards he repurchased the mansion from
Forde Hughes but he did not immediately occupy it and for some
three years it was the home of a retired clergyman. Some time later
Lloyd Williams assembled an imported Scandinavian bungalow
on a new forty-acre holding that he formed out of the lands of the
mansion and home farm. He occupied the bungalow and farmed the
holding, his social contacts being virtually limited to occasional
visits to other gentry families in the neighbourhood. Eventually he
returned to the mansion where he lived until his death during the
Second World War. He was the last to live in Gwernant, which was
a plain but well-proportioned and attractive Georgian structure.
Its roof has subsequently been stripped and it is now a ruin.

The other mansion that stood in the parish was Troedyraur
House, once the seat of the Bowens, who were connected with one
of the best-known families in Wales, namely the Bowens of Llwyngwair (in Pembrokeshire). The Revd. Thomas Bowen, Rector of
Troedyraur, who died in 1842, was the last of the line of Troedyraur
House. He was the grandson of John Bowen, also of Troedyraur,
who contributed to the Welsh literary life of South Cardiganshire
during the eighteenth century, composing among other things
halsingod, a form of carol characteristic of the three south-western
counties of Wales during the seventeenth and first half of the
eighteenth centuries. It was not at all uncommon that the minor
gentry of the area were either patrons of, or contributors to, the
Welsh literary life of the region during that period.

The Revd. Thomas Bowen was possessed of a small estate of three
farms (including the home farm) and five small holdings, totalling
538 acres. Among the many squires who actively promoted the
advancement of agriculture in this area during the late eighteenth
and early nineteenth centuries he gained for himself a reputation as
an 'enlightened and successful agriculturalist'. In 1842 the tenure
of the estate passed to James Bowen (1806–72), High Sheriff
of Cardiganshire in 1848 and the second son of James Bowen of

---

[11] *Cardigan and Tivy-Side Advertiser*, 15 May 1914.

Llwyngwair. On his death the estate passed to his son the Revd. A. J. Bowen (1839–1920) who had married the daughter of Aeneas Cannon, M.D., of Cheltenham, and was rarely resident as he at various times held livings in Somerset (1874–91), Berkshire (1896), and Dorset (from 1896 until his retirement). In A. J. Bowen's absence one Frederick Corbyn, Surgeon-General, Her Majesty's Indian Service, came into residence, and in 1882 his daughter Florence Emma was married to George Bevan Bowen of Llwyngwair. It appears that the mansion was later rented to Forde Hughes, Esq., for a short period before Eynon George Bowen, the owner's son, came to reside there, having married Georgina Catherine Wills of Clifton in 1890. He continued there until about 1900 when he removed to Clifton where he died in 1916, to be survived by his father, the Revd. A. J. Bowen.

Before his departure from Troedyraur Eynon Bowen occupied the mansion and home farm of 177 acres. He also farmed another property of 111 acres where he installed a bailiff and became known locally as a Shorthorn breeder. (This was of consequence because Shorthorns matured earlier than the native Welsh Blacks, and the English winter beef producers to whom they were ultimately sold favoured the earlier maturing store cattle.)[12] But he ceased farming during the 1890s and before the end of the century the demesne was rented as grazing land. He himself retained only sufficient land to keep hunters and coach horses, and land to keep milk cows to provide for the needs of the house. Meanwhile he was Master of the Tivyside Fox Hounds and on occasion the secretary of the Tivyside Agricultural Society which was then patronized by the gentry and was responsible for the annual show held at Newcastle Emlyn, the centre of the Tivyside gentry's social occasions. He was also a partner in a firm of auctioneers in Newcastle Emlyn, but being unable to speak Welsh his circle and his influence were inevitably more restricted than might otherwise have been the case.

After his son's departure A. J. Bowen disposed of the greater part of the estate which was reduced to the mansion and the home farm. This was taken in hand and pedigree Shorthorns were bred by the next tenant of Troedyraur House, P. Cecil Phillips, Esq. This lasted until 1912, when the Revd. A. J. Bowen returned to the mansion for a time, letting the home farm to local tenants. This arrangement

[12] For this matter in general see R. L. Jones, 'Changes in the Pattern of Cardiganshire Farming', *Journal of the Royal Welsh Agricultural Society*, vol. xxvii (1958), pp. 39–50.

continued until the sale of the property in 1921, which saw the end of the house as a country seat of the squirearchy. In the case of Troedyraur House and of Gwernant we note the smallness of the estates and the succession of tenants. In the case of Gwernant the mansion was for the greater part of forty years little more than a place of retirement for army officers, without estate in the area, without even a home farm, with few local ties or responsibilities, and unfamiliar with the language of the overwhelming mass of the populace. When the owner came he did so as a stranger, to occupy the lodge while the mansion passed to an eccentric. In the case of Troedyraur House we note the reduction of a small estate into a home farm which was eventually let as grazing land during a period of major changes in agriculture. The owner was seldom resident and was a squire who did not farm, though the area was overwhelmingly agricultural.

In the adjoining parish of Llangynllo stood the mansion of Mount Gernos, the home of the ancient line of Lewes which came to an end through lack of a male heir during the late seventeenth century. Thereupon the estate passed to the Parrys on the marriage of the heiress of Mount Gernos to Thomas Parry, Esq., of Cwm Cynon, Llandysiliogogo, in South Cardiganshire. During the middle of the nineteenth century a similar situation occurred when Thomas Llewelyn Parry's only child was a daughter, Judith. She married Gwinnet Tyler, Lieutenant, R.N., in 1852. Gwinnet Tyler was the brother of Lt.-Col. George Henry Tyler of Cottrell, Glamorgan, and they were the sons of Vice-Admiral Sir George Tyler of Cottrell, who had married Harriet Margaret, the daughter of the Rt. Hon. John Sullivan of Ritchings, Berkshire.

The estate of Mount Gernos comprised 1,373 acres in Llangynllo parish with a further 282 acres in the parish of Llandysiliogogo (with a total rental value of £963. 12s. 10d. in 1907).[13] The estate was, however, encumbered with the debt of Thomas Llewelyn Parry, but on the death of Lt.-Col. George Henry Tyler in 1878 a life interest in his estate was inherited by Gwinnet Tyler, Esq., of Mount Gernos. The mansion stood on the western side of a narrow and particularly steep-sided valley fed by a stream draining into the river Teifi. During the years 1880 and 1881 Gwinnet Tyler built a woollen mill in this valley, with a hall and reading-room for the

[13] Mount Gernos Estate Catalogue in N.L.W. Other details of the Mount Gernos Estate are to be found in the Morgan Richardson Collection in N.L.W.

use of employees, and for others who lived in the locality. Gwinnet and Judith Tyler had three sons and three daughters: two of the sons were sent to Leeds and other textile centres in Yorkshire for training in the textile industry, and they managed the factory as long as it remained the property of the family. Of the six children only one, a son, married, the others living in Mount Gernos until the family's connection with the estate came to an end in 1918.

One of the sons had been intended for the Church, and the Tylers were devout churchmen who had been influenced by the Oxford Movement. They had built a chapel in the attic of Mount Gernos where evening services were held, and a Sunday school was conducted in the mansion, in the English language, for the children of estate workers and others. After their coach had been sold they continued for years to walk a long and difficult road to the parish church. Some members of the family had a slight understanding of Welsh, but no more.

In 1884 Gwinnet Tyler found it necessary to raise a mortgage of £20,000 on the estate, and the household's income was further depleted when the money that Gwinnet Tyler received personally from his life interest in his brother's estate ceased on his (Gwinnet Tyler's) death in 1886. The family farmed the home farm until 1901 when it was let to a tenant. But in 1907 with £18,000 still outstanding on the mortgage it became necessary to sell the greater part of the estate which was thereafter reduced to the mansion, home farm, and two small holdings, totalling 192 acres, along with the woollen mill. The mill was sold in 1916 and the rest of the property two years later. The mansion and home farm were purchased by a South Cardiganshire business man for £5,500, and in the following years the mansion was extensively modernized at a cost of £7,500. The post-war slump affected the owner's fortunes and in 1922 the property was again offered for sale. The mansion and home farm were sold separately, the former being purchased by a farmer who was also a livestock dealer for the sum of £900. The new owner stripped the mansion of its fittings and eventually converted the shell into a piggery, which it remained until 1960.

But the most influential family in the neighbourhood were the Lloyds of Bronwydd, which, like Mount Gernos, stood in the parish of Llangynllo. While the expression 'great people (*gwŷr mowr*) referred to all members of the gentry, the Lloyds were spoken of as 'very great people' (*gwŷr mowr iawn*). The most concise way

of describing the regard in which they were held and the prevalent
ideas about them is by quoting the obituary notice of the last
member of the family in the male line, and the last to reside at
Bronwydd where he died in 1933;

Sir Marteine Owen Mowbray Lloyd, of Bronwydd, Cardiganshire
and Newport Castle, Pembrokeshire, Lord of the Barony of Kemes,
died on Tuesday at the age of 82.

The only son of the first baronet, whom he succeeded in 1877, he
married in 1878 Miss Katherine Dennistoun, of Golfhill, Lanarkshire.
Sir Marteine was a lineal descendant of Martin de Tours and seventeenth in direct descent from Thomas Plantagenet, Earl of Norfolk,
eldest son of Edward I of England, while Lady Lloyd was descended
from Robert II of Scotland. Sir Marteine also traced back to the Lords
of Kemes and Bronwydd in the ninth century, held the same lands
that his ancestor held in 1241, and was patron of the same six livings.
One of his rents was a red rose, a custom dating from the time of Henry
VI.

Sir Marteine was the only Lord Marcher in the Kingdom, and possessed the unique right, which he exercised, of appointing the Mayor
of Newport, Pembrokeshire, every year. He was High Sheriff of
Cardiganshire in 1881, and was Deputy Lieutenant and Justice of the
Peace for the counties of Cardigan, Pembroke, and Carmarthen. From
1874 to 1887 he held a captaincy in the Pembrokeshire Yeomanry, and
was a popular figure in hunting and sporting circles in West Wales.
Four years ago he and Lady Lloyd celebrated their golden wedding,
when he presented the Mayor and Court Leet of Newport, Pembrokeshire, with a gold mace, and in turn received a set of gold plate from
the Mayor and burgesses and other gifts from the county magistrates,
the tenantry, and others. . . . There is no heir to the baronetcy, his only
son, Captain Kemes Arundel Lloyd, Grenadier Guards, having been
killed in action on the Somme in 1916. There are three daughters.[14]

Soon after her husband's death Lady Lloyd removed from Bronwydd. In 1937 the mansion, an ornate Victorian Gothic structure,
along with that part of the estate (comprising 2,072 acres) situate
in South Cardiganshire were offered for sale. Since then the roof
has been stripped and Bronwydd is now a ruin.

The Lloyds, who had long been resident at Bronwydd, were
noted for their generosity and for their support of churches and nonconformist chapels in the area. Colonel Thomas Lloyd (d. 1807)
and his wife were evangelically inclined; in 1794 he built a
Methodist chapel (Capel Drindod, lit. Trinity Chapel) on his own

---

[14] *The Times* (London), 6 Apr. 1933. Reproduced from *The Times* with permission. It may be remarked that the factual accuracy of several of the statements
in this notice is doubtful.

land and it was transferred to the Calvinistic Methodists when they were constituted a separate body. He had endowed the chapel with £600 and the interest thereon, and later his widow added a similar endowment. He defrayed the cost of rebuilding the parish church (of Llangynllo) and provided an endowment for the ecclesiastical chapel of Capel Cynon. His wife, the daughter of John Jones, M.D., of Haverfordwest, had in 1796 assigned nine turnpike tallies to the 'value of £300, in trust for the minister of the congregation of the protestant dissenters meeting in St. Thomas' Green . . . in Haverfordwest . . . in compliance with the request of Sarah Jones, deceased',[15] her mother.

Capel Drindod, as built by Col. Thomas Lloyd, was an unusual structure having three galleries each with its own staircase and doorway in the gable end facing the pulpit. The central gallery was that of the Bronwydd family, the one on the right (viewed facing the pulpit) that of the youths, and the gallery on the left was that of the young women. Like Col. Thomas Lloyd his son (Thomas Lloyd, 1788–1845) worshipped at the parish church in the morning and at Capel Drindod in the afternoon. Thomas Lloyd stayed for the singing practice that followed the service, keeping time for the singers and whistling the tune. During his lifetime the estate band provided the accompaniment for the services in the chapel. He was the last member of the Bronwydd family to worship regularly at Capel Drindod but sympathy for church and chapel characterized the family while they remained at Bronwydd.[16]

Sir Thomas Lloyd (1823–77) and Gwinnet Tyler, Esq., of Mount Gernos, were together responsible for rebuilding the parish church in 1870. When Bryngwenith Congregational chapel was rebuilt in 1882, Sir Marteine Lloyd provided the land and his daughter laid the foundation stone. In fact, this sympathy towards nonconformists was by no means rare in the area. Thomas Llewelyn Parry, the last of that name to occupy Mount Gernos, provided land for and contributed towards the building of Bwlchygroes Congregational chapel which was opened in 1833, inviting those ministers present at the opening ceremony to dinner at Mount Gernos. In 1865 the Walters of Glanmedeni built and endowed the Calvinistic Methodist chapel named Watch Tower.

Sir Marteine Lloyd occasionally attended evening service at

Capel Drindod and at the Congregational chapel of Bryngwenith; he attended funeral services of old family servants at Gwernllwyn, another Congregational chapel, and when invited he chaired various meetings and lectures (in Welsh) at the last-named chapel. Nonconformist ministers were received at Bronwydd on the same terms as clergymen, and for some years Sir Marteine Lloyd let the small mansion of Kilrhue (in North Pembrokeshire) rent free as a residence for the local Congregational minister. The living of Llangynllo was in the gift of the freeholders of the parish. Sir Marteine Lloyd was the most influential member of the appointing committee which merits mention when we recall that the incumbent was an active revivalist during the religious revival of 1904 and 1905. The living of the parish of Troedyraur was in the gift of the Lord Chancellor, and was held for many years by the Revd. Rhys Jones Lloyd (1827–1904), the younger brother of Sir Thomas Lloyd (1823–77) and the uncle of Sir Marteine Lloyd.

Like his nephew, Rhys Lloyd was a keen sportsman, keeping hunters at the rectory and riding to hounds till he was in advanced age. For many years he was the honorary secretary of the Tivyside Lawn Tennis Club. But he was best known as a generous parish priest who considered churchmen and nonconformists alike to be his parishioners. He frequented such especial meetings as the harvest festivals of all denominations, and was present and helped to officiate at nonconformist burial services. In a period when there was much antagonism between the supporters of the British and the National Schools Societies in the country at large he chaired meetings held in the Congregational chapel to raise funds for a British school in the locality. His coachman was a faithful Congregationalist who was regularly released from his duties on the appropriate evenings in order to attend meetings at the Congregational chapel. At the height of the anti-tithe agitation of the late nineteenth century he regularly helped certain ailing nonconformists who were known for their fierce opposition to the payment of tithes. A man of wide literary interests, he was much more familiar with the literature that engaged the educated gentry at large than with the Welsh literary revival that began during his lifetime.

Sir Marteine Lloyd farmed Bronwydd's home farm until the year 1885 or thereabouts, thereafter sufficient land was retained to keep horses and the rest let as grazing land. Poultry and dairy produce were purchased from the tenants. Timber and building-stone were freely available to the tenantry from the estate's resources, tenants being shown into the drawing-room to make their requests personally. Indeed it was considered an honour to be a tenant on the Bronwydd estate, chiefly because the Lloyds treated others not only with generosity but with respect. The attitude they engendered in others is best conveyed by what so many of their one-time tenants remarked, 'They were very great people, but they were not snobs.'

They would give lifts in the family coach to tenants carrying loads, and Sir Marteine though a particularly keen sportsman would not hunt over other people's land without their invitation. This was very widely extended. Tenants themselves reared hounds for the Bronwydd Beagle pack and poaching was the one unpardoned offence on the estate. Sir Marteine was an active member of the committee of the United Counties Hunters Society, and took a close interest in the United Counties Agricultural Society. A man of exceptional generosity he held, among many other offices, the Presidency of the West Wales Discharged Prisoners Aid Society and was chairman of its working committee.

As has been stated, Sir Marteine Lloyd occasionally took the chair at meetings conducted in Welsh and many accounts remain current of his confusions when speaking a language in which he was by no means thoroughly familiar. While delivering the chairman's address at Gwernllwyn chapel when Cranogwen (a formidable lady who was nationally known at poet, preacher, and temperance worker) was the lecturer, he stated that though he had frequently chaired meetings for a 'male man' (*dyn gwryw*), a common local expression, that was the first time he had chaired a meeting addressed by a 'female man' (*dyn menyw*), the logical opposite of *dyn gwryw* but idiomatically incorrect. In tragic vein, he addressed men emerging from Bryngwenith chapel on an evening during the Great War (in which he was to lose his only son) inviting them to come and be killed ('*Chi dod i gal y'ch lladd*') confusing the word *ymladd* (to fight) with *lladd* (to kill). But the Lloyds are remembered for their generosity above all else, and their readiness to help whenever help was needed. And this despite the fact that when Sir Marteine Lloyd succeeded to the estate in 1877 it was mortgaged to the extent of £94,000. By 1922 this sum had been reduced to £24,000 through the sale of various properties during the intervening years.[17]

[17] N.L.W., MS. Bronwydd 5995; see letter to Sir Marteine Lloyd, dated 5 Jan. 1922; see letter from Sir Marteine Lloyd dated 19 Sept. 1922.

Details have been noted above of those gentry with whom people in the parish of Troedyraur were in closest contact, as landlords, employers of labour, as worshippers at the parish church and occasionally at nonconformist chapels, on whom people called when collecting for charities, and who extended charity to those whose needs were brought to their notice. But their influence was not confined to the few families living in the locality for the gentry occupied statuses in two social fields: they were at one and the same time the top stratum in the threefold structure of local society (p. 13) and also members of a nation-wide stratum of British society, sharing in that stratum's general characteristics. Here are noted some further details that bear on the gentry's membership of that particular and distinctive stratum, and some of the material mentioned above is recapitulated to this end.

It was (as stated) one John Lloyd Williams who at the beginning of the nineteenth century built Gwernant, the mansion whose ruins now stand on the slope overlooking the vale of Troedyraur. His third son, Henry Lloyd Williams, was an officer in the Honourable East India Company's Bengal Army, taking holy orders on retiring from that army and becoming a chaplain in India. Henry Lloyd Williams's son was a captain in Her Majesty's 24th Regiment, serving in India, where he was badly wounded. He in turn had one son who survived, Gilbert Lloyd Williams (who later returned to Gwernant), a daughter who married a government engineer in Ceylon, and another daughter who was married first to a planter in Ceylon and then to a captain of the Army Pay Department.

On the death of John Lloyd Williams's grandson in the direct male line without issue in 1870 a life interest in the estate passed to his widow, who removed from Gwernant and survived until 1898, as has been explained above. During this time Gwernant was let to a number of tenants, and it has been noted that the last of them (before Gilbert Lloyd Williams came into the estate) was a retired army officer James Graham Kell. One of his daughters married in Cape Town, South Africa; another married a lieutenant in the Essex Regiment. He was the cousin of Eynon Bowen of Troedyraur House, whose brother was a captain in the Kimberley Light Horse. Troedyraur House was owned by a clergyman, the Revd. A. J. Bowen, whose brother was a lieutenant-colonel in the 88th (Connaught Rangers) Regiment, and at one time, while the Revd. A. J. Bowen held a living in southern England, Troedyraur House was (as has been mentioned) let to one Frederick Corbyn, Surgeon-General, Her Majesty's Indian Service. When his daughter married George Bevan Bowen of Llwyngwair, a lieutenant in the Pembrokeshire Yeomanry, they remained in regular correspondence with those Corbyns who remained in India. Forde Hughes, Esq., educated at Bath and Foccy (in France), who at some time rented Troedyraur House and purchased Gwernant, was a first lieutenant in the Royal Glamorgan Artillery Militia.

Gwinnet Tyler, at nearby Gernos, had served as a lieutenant in H.M.S. *Leander*; his brother was a lieutenant-colonel who served in the Crimean War and in India. They were the sons of a vice-admiral who had married the daughter of a Berkshire squire. Their uncle, the son of an admiral, was Rector of Llantrithyd (Glamorgan) and Vicar of Mynachlogddu (Pembrokeshire). Sir Marteine Lloyd of Bronwydd was a captain in the Pembrokeshire Yeomanry, his mother was the daughter of George Reid, Esq., of Bunker's Hill and Friendship estates, Jamaica, and of Watlington Hall, Norfolk. One of his father's brothers was a captain in the Warwickshire Regiment, another of his father's brothers was rector of Troedyraur. He himself had married into a Lanarkshire family, spending part of the year in Scotland, staying in London during the 'season', and residing at Newport Castle when away from the family seat at Bronwydd which was manned by a skeleton staff for much of the year.

These details, along with those given earlier, show that the gentry families of the locality were members not so much of a nation-wide as an empire-wide stratum of society, closely connected with the armed forces on the one hand and the Church on the other. They shared the sentiments generally associated with that stratum of society, being strong supporters of the monarchy and the Established Church despite their sympathy for nonconformists (rather than nonconformity), believers in the value of empire, and in people's duty to provide armed support for Crown and Country. In much of this they differed markedly from another leadership that had arisen, namely that provided by nonconformist ministers and prominent laymen, as shall be discussed below.

In the local society wherein they constituted the top stratum, they were, with the importance that they attached to blood and descent, an exclusive group, marrying within their own social stratum whether locally or elsewhere within Britain and the Empire. In the local agricultural system they were the landlords whose tenants almost always held on a yearly basis and were thus liable to receive notice, as some did for game offences. Otherwise notice was very rarely given. Again as landlords they were responsible for the supply of fixed capital while their tenants provided the working capital. When their financial resources were reduced, whether for personal or more general reasons (as the decline in the value of land during the agricultural depression that began during the 1870s), there was no adequate alternative supply of capital till the government acted during the Second World War. When farms were sold tenants purchased them though they lacked adequate resources to provide both the fixed and working capital, as these examples show:

Farm 1. 193 acres: purchased in 1881 for £2,275. Of this £1,650 was borrowed, at 3¾ per cent. By 1895 none of the capital had been repaid.

Farm 2. 146 acres (including moorland): purchased in 1889 for £825. Of this £600 was borrowed at 3¾ per cent. By 1895 none of the capital had been repaid.

Farm 3. 52 acres: purchased in 1870 for £4,000, all of which was borrowed at 4 per cent.

Farm 4. 120 acres: purchased in 1883 for £2,250, of which £2,200 had been borrowed, at 4 per cent.

Farm 5. 95 acres: purchased in 1885 for £2,040, of which £2,000 was borrowed at 4 per cent. By 1895, not only had no repayment been effected but the purchaser had borrowed another £500.

Farm 6. 110 acres: purchased in 1879 for £4,150. Of this £2,000 was borrowed at 4 per cent. By 1895 nothing had been repaid.[18]

The first concern of such farmers was the repayment of debts, and quite commonly there was little improvement of fixed capital (other than the erection of corrugated iron haysheds) until the Second World War.

The gentry were not only landlords and employers of labour, they filled those offices that carried prestige on a national level, as Deputy Lieutenants and High Sheriffs, and the squires of Gwernant, Troedyraur House, Gernos, and Bronwydd along with the Revd. Rhys Jones Lloyd all sat on the local bench.

But while they were members of a nation-wide stratum of society

[18] N.L.W., MS. D. Lleufer Thomas 3602E. Re the variation in prices note that one farm was bought before the beginning of the agricultural depression which started in 1879; note too that some of the farms included moorland.

and at the same time constituted the 'apex' of local society the class division between them and the rest of the society was complicated by the fact that in respect of literary culture they (the gentry) shared in the characteristics only of that national stratum to which they belonged. They had no place (as patrons, contributors, or participants) in the literary culture familiar to the extent that any culture was familiar to most members of the local society. Redfield has used the notions of a 'great tradition' and a 'little tradition' to elucidate the relationship between the culture of a local society (the little tradition) and the main stream of a nation's culture (the great tradition). 'The culture of a peasant community . . . is not autonomous. It is an aspect or dimension of the civilization of which it is a part . . . peasant culture requires continual communication to the local community of thought originating outside of it. . . . In a civilization there is a great tradition of the reflective few, and there is a little tradition, of the largely unreflective many.'[19]

But in the area considered in this study (as in others) there were two great traditions, that of English literary culture in which the gentry in varying measure shared, and that of Welsh literary culture, in which the gentry, through ignorance of or incomplete grasp of the language, did not share. But the 'little tradition' of this locality was Welsh. Two matters need to be mentioned: first that this 'little tradition' was a living tradition, second that the gentry could not link this 'little tradition' with its corresponding 'great tradition'. There is little difficulty in illustrating the first point, while the second is self-evident. People live their lives within a cultural tradition as well as in a particular society, and in respect of cultural tradition the gentry, the social leaders, could provide no effective leadership.

In the later seventeenth and early eighteenth centuries the minor gentry were patrons of and contributed to the literary activity of South Cardiganshire and the Teifi Valley generally. The Lewes family of Gernos were patrons while Erasmus Lewes (d. 1745), educated at Jesus College, Oxford, was an active contributor. John Bowen, Esq., grandfather of the Revd. Thomas Bowen (d. 1842) of Troedyraur House, wrote a type of verse that was characteristic of south-west Wales in his day. Theophilus Evans (1694–1769) of Penwenallt (which at one period of the nineteenth century was occupied by a branch of the Lloyd Williams family of Gwernant) saw certain of the manuscripts on which he based his classic *Drych*

---

[19] R. Redfield, *Peasant Society and Culture* (Chicago, 1956), pp. 68–70.

y Prif Oesoedd in Llwynderw, the home of William Lewes, Esq. Stephen Parry, Esq. (d. 1724), of Neuadd Trefawr and Walter Lloyd, Esq., of Coedmor were both patrons of a volume of translations from the English made by the Revd. Alban Thomas (*c.* 1650–1734) of Blaen-porth, himself of gentle birth, whose other work provided one of the earliest titles to be printed on the first printing press to be established in Wales, at Atpar, Newcastle Emlyn, in 1718. From these men's plebeian contemporaries (e.g. Ifan Gruffydd, Twrgwyn, (*c.* 1655–1735), and Siencyn Thomas of Cwmdu (1690–1762)), a continuous chain of culture-bearers can be traced down to the present day.

Herewith an example is provided. Siencyn Thomas (1690–1762) was a shoemaker who had mastered the Welsh restricted verse measures. He was survived by three sons who wrote poetry. Of these Sion Siencyn (1716–96), also a shoemaker, composed verses to the gentry and frequented the competitive meetings (*eisteddfodau*) of his day. His brother Nathaniel Siencyn (1722–99) sang to *Palas Troed yr Aur*, that is, to the Bowens of Troedyraur House. A South Cardiganshire poet, the Revd. David Davies (1743–1826) of Castell Hywel, officiated at Nathaniel's funeral, and when David Davies published a volume of verse a prefatory poem was contributed by the Revd. Daniel Evans (1792–1846) to whom David Davies paid thanks for help in completing the work. The Revd. Daniel Evans (*Daniel Ddu*) provides a clear example of the linkage between the 'little tradition' of regional literary activity and the 'great tradition' of the main stream of national literary life. A fellow of Jesus College, Oxford, he was at one and the same time a bardic teacher adjudicating and teaching at minor competitive meetings in South Cardiganshire (as at the Alltyrodyn Arms in 1823) and a prominent figure in the establishment of the competitive meeting (*eisteddfod*) in its modern form. He was in contact with and for a time one of the major figures of the day in national literary life. He was also one of a number of local poets who addressed one another in verse. Among them was Evan Thomas (1795–1867), *Bardd Horeb*, a tailor who had been taught by the Revd. Daniel Evans. He sang to Thomas Lloyd, Esq., of Bronwydd, the last of the family to worship regularly at Capel Drindod, and to John Jones (Ioan Cunllo, 1806–84), a shopkeeper poet of Rhydlewis, in Troedyraur parish. John Jones himself was a personal friend of both Evan Thomas and the Revd. Daniel Evans and at the Lampeter *eisteddfod* of 1859 he won the

prize offered for an elegy in the latter's memory. By the time of
John Jones's death the local *eisteddfodau* were very firmly established. For generations they provided training in the mastery of
verse in the restricted measures through the medium of the detailed
criticism delivered by the *eisteddfod* adjudicators. Thus they have
schooled bearers of this tradition who are active at the present day.

As members of a nation-wide stratum of society with the high
standards of literacy, material goods, and hygiene characteristic
of that stratum, the gentry did constitute a linkage whereby alternative standards, some higher than and some different from those
commonly accepted, were presented to other members of the local
society. They provided the lead that was necessary to establish the
first District Nursing Associations in the area, different families
opening their gardens in turn in order to raise funds, contributing
substantially themselves, and organizing collections to that end.
Similarly they attempted to raise standards of elementary hygiene
more particularly among their own tenantry whom they could best
influence. The mansions were, too, centres for the diffusion of the
English language as English was the language of every gentry
family. This influence was most obvious not among their tenantry
but among the professional people that the gentry employed,
lawyers, land-agents, doctors (who would in any case be fluent in
English).

But the reaction that this produced in others is best seen in the
notion of *gwerinwr*, literally 'a man of the commonalty'. The first
known usage of the word dates to the sixteenth century, but thereafter its use is unknown until 1834.[20] It became popular during the
nineteenth century. By *gwerinwr* was (and is) meant a man who
consciously maintained a position that was seen to be in contra-
distinction to that of the gentry, perhaps a man who though in a
position to emulate the gentry, chose not to do so, but identified
himself with the commonalty and participated in those activities
that characterized the commonalty rather than the gentry. One
example must suffice. J—— D——, a shoemaker's son, retired at
the age of forty, having amassed a fortune in business in an
industrial area in South Wales. He then returned to his native village
where he built himself an imposing house. He was always well
dressed and had the necessary means to associate with the gentry.
Instead he associated with those of that section of society that he

[20] Information from the editor of *Geiriadur Prifysgol Cymru.*

belonged to before he had become a successful business man. He spoke Welsh fluently and habitually, he took a prominent part in the affairs of the local chapel where he was a deacon. As a man of unusual intelligence he was a local leader whose exceptional status was clear to all. But he remained a *gwerinwr*, widely read, able to participate in the 'little tradition' and to be a link between it and the main stream of national culture. This the gentry could not do, hence the need of and opportunity for another leadership. This was a situation where the content of culture was of immediate sociological relevance, for it was in part the content of the gentry's culture that disabled them from constituting the medium whereby the main stream of national culture was presented to the other members of the local society that they together comprised.

As the nineteenth century drew to its close other factors affected the gentry's position in local society. Successive extensions of the franchise, reforms of local government and of the bench, which gave opportunities to many who had not previously had such opportunities, the relative impoverishment of landowners with the decline in land values and the fall in prices of agricultural products that came with the agricultural depression during the last quarter of the nineteenth century consequent upon the introduction of refrigerated meat ships and the cultivation of American prairie grainlands, the incidence of death duties; all contributed to a situation wherein leadership in local society was more difficult and less attractive to the gentry than it had been. Those without home farms could contribute little to the agricultural changes that were occurring, in marked contrast to the role of the gentry in the agricultural changes that had occurred a hundred years earlier. Sporadic selling of outlying parts of estates began around 1875, but the period that virtually eliminated the estates were the years immediately before and after the First World War. In 1910 the gentry owned fifty properties in Troedyraur parish; in 1921 they owned eighteen.

NOTE
(*see p. 2*)

Herewith an account of the major documentary sources on which are based statements and tables to be found in the body of this study.

(*a*) Census of England and Wales. Apart from population statistics these are valuable because they state the number of occupied houses in each parish. This being so it is possible to know during the course of investigation the percentage of households for which one does or does not have information. Further, after 100 years have elapsed it is possible to consult (at the Public Record Office) the 'recorders' notebooks' wherein were recorded the details on which the published census data were based. These state according to households the name, age, and marital state of every person resident in any parish. The relevant 'recorders' notebooks' for 1861 have been consulted.

(b) Electoral lists. These contain the names of all houses wherein resided people qualified to vote, and the names of all the people who were qualified to vote in both central and local government elections. Changes in the occupation of houses and farms can be found accurately from perusing successive electoral lists, as well as the dates of such changes. Further, during the late nineteenth century the electoral lists distinguished between 'ownership electors' and 'occupation electors' the former qualifying as electors because they owned property and the latter qualifying because they occupied property. It is thus possible to distinguish between tenants and freeholders. The place of residence as well as the name of the qualifying property is given in each case.

As the number of occupied houses is known (from the details contained in the Census of England and Wales), a consideration of the data provided in electoral lists enables one to say that these lists virtually state the names of all dwellings and of all householders. This is of importance in the course of investigation because it enables the investigator to ask specific questions about houses and householders rather than general questions, as for instance when inquiring which farms formed co-operating groups for certain agricultural operations.

(c) Directories. E.g. *Kelly's Directory of South Wales* for various years, as, for instance, 1891, 1906, 1914. This work gives for each parish the names of the 'principal landowners', of commercial establishments, and of each farmer along with the name of his farm.

(d) Sales catalogues, which were published when estates were sold. The catalogues give details (and frequently maps) of each farm, including acreage, name of tenant, and rent. Where the catalogues contain maps these show the names of the owners of lands adjoining the properties offered for sale.

(e) Copies of ploughing assessments made during the First World War also state the size of holdings at the time when the assessments were made.

(f) The annual reports of various chapels. These were printed as from the early years of this century and contain the names and addresses of all members, arranged by households.

(g) School registers. The Cardiganshire Local Education Authority has placed various school registers in the National Library of Wales. They include 'Admission, Progress, and Withdrawal' registers covering the period 1872–1920 in the area with which this study is chiefly concerned. In them are to be found the names of children and of their parents (or guardians), their addresses, dates of birth, dates of admission to school, dates of departure from school, which schools (if any) children had previously attended, indicating when and from where a household had moved into the district. In addition information is given if children entered secondary schools on their departure from primary school.

It would not be possible to present the details and tables given in Chapters VI and VII especially, were it not for the information contained in these sources. On the other hand these sources would have remained confusing were it not possible to make personal inquiries. In conjunction these procedures have provided the data on which is based much of the material presented herein.

II

FARM PRACTICE AND SOCIAL STRUCTURE

I

BEFORE they started to sell off their estates, twelve gentry landlords owned 2,400 acres in the parish of Troedyraur, which had a
total area of 4,064 acres. The remainder was owned by common
people of whom many, but not all, farmed their own holdings. It is
now time to discuss some of the considerations relevant to the agricultural system early in the period with which this study is concerned, which takes us back to the last decades of the nineteenth
century.

It has been seen that climatic conditions render the whole area
more suitable for grass production and stock rearing than for
arable farming. The character of farming in the area was affected
too by South Cardiganshire's relative remoteness from urban and
industrial markets before the development of motor transport. A
farm's cash income came chiefly from the sale of store cattle reared
from the farm's own stock, from the sale of pigs to dealers (either
at the farm or in the nearest market town), and from the sale of
casked butter to butter merchants at the market towns. As there
was a shortage of winter feed, and under the environmental conditions obtaining, cows did not become ready to be served by a bull
until June of each year. They then calved in the spring and the
calves could graze during the months when there was keep on the
ground. The cows thus gave their milk chiefly in summer, with a
marked fall in supply during the winter.

Very little fresh milk was sold, for there was little available during
the winter, while the summer milk was needed for butter making.
(Daily milk deliveries did not start in Newcastle Emlyn till 1900
or thereabouts.) Very little butter was sold in one- or two-pound
pats, for most of it was heavily salted and packed into tubs of
twenty pounds to thirty pounds each, or into casks of 100 pounds
to 120 pounds each. These were taken to the market towns for sale to
the dealers. The sale of bulk milk did not begin until the 1920s when
a creamery was established at Capel Dewi (in South Cardiganshire);

it became of major importance only after the establishment of the Milk Marketing Board in 1933.

Artificial feeding-stuffs were little used. The account books of one of the most important South Cardiganshire merchants then engaged in the coastal shipping trade show the importation of very small quantities of linseed cake and nothing else.[1] In addition Indian corn was purchased, to be fed regularly to the poultry, and ground to feed the pigs and to supplement the cattle's food in periods of scarcity. Hence it was necessary to grow almost all the feed on the farms, both the required winter feed and feed for fattening pigs before they were slaughtered or sold. Similarly most of the household foods were produced locally. Bread corn, oatmeal, potatoes, bacon, cheese, and vegetables were the most important.

Stock-rearing makes relatively constant demands on labour. More stock could be carried in summer when natural feed was available and it was then that the duties of milking and butter-making were heaviest. But though these duties were lighter in winter it was then that the animals were indoors at night and natural feed was insufficient. Thus it was necessary to prepare their feed and clean the byres daily. On the other hand, crop cultivation makes heavy seasonal demands on labour, and crops had to be grown because little imported foods, artificial or otherwise, were available to men or animals. Wheat was grown for bread corn (finally displacing barley in this area during the 1880s and 1890s), oats to provide oatmeal for the house and feed for the horses, whose feed was supplemented by gorse, a protein-rich crop grown in plantations. Barley was fed to cattle and pigs along with hay (to the cattle) and root crops. In Troedyraur parish in 1900, 64 per cent of the holdings were of less than 30 acres, 75 per cent were less than 65 acres, and 150 acres was considered a large farm (detailed information is given in Table 9, p. 269). Such holdings, when chiefly dependent on store cattle and casked butter, gave but poor financial returns:

. . . a smallholder can never make much of a living out of livestock alone; the turnover is too slow and his own labour contributes too slight a share to the value of the product. Of course he may do well enough by producing and selling milk, because he adds the routine of a milk round to the work of feeding and milking his stock . . . Butter making and calf raising also bring in regular returns but do not require so much labour, hence the profits to the smallholder are less. . . . The slowest return, because the minimum amount of labour is

[1] W. Jones, *Deilad y Dole,* unpublished typescript in private possession.

expended on the product, comes from rearing stock on grass until they are fit to sell as stores; in itself the operation is profitable enough, but the profit is small per acre. . . .[2]

So commented an authority on British agriculture, Sir A. D. Hall, in 1910. As a result farms could only be staffed to meet their routine needs, depending on other arrangements for unusual and seasonal labour demands, though these were of major importance.[3]

While it would be quite incorrect to describe the economy as a subsistence economy for the farms depended on the sale of stock and dairy produce, yet the details given above show that there was a much greater dependence on cultivation for direct consumption, and of a subsistence element in the economy, than is now the case. Witness such facts as that straw thatching-ropes were made on the farms as were the pegs used for thatching. Boot blacking was made from pig fat and soot, while pigs' bladders were kept for storing lard. When calves were slaughtered the stomachs were kept in order to prepare rennet for cheese making. Candles were still made in many homes. Just as the farms depended on their own resources for feeding-stuffs, they did so too for manures. Only two artificial fertilizers were used, and those in very small quantities, super-phosphates and dissolved bone. During the 1880s an intelligent man who was considered an advanced farmer used five hundred-weights of superphosphates and three and a quarter hundredweights of dissolved bone, per annum, on his forty-acre holding, which would have been a liberal amount for two or three acres.[4] Guano came into use about 1900, and basic slag shortly afterwards. (Guano and superphosphates became known in Britain during the 1840s, the first cargo of guano having been imported in 1835.)[5]

Lime was used extensively. Otherwise the only fertilizer used in quantity was farmyard manure. This accumulated while the animals were stall fed on the farm's own feed during the winter months. But winter feed was frequently insufficient in quantity, on the smaller holdings calves often died in spring if the growth was late. Much of the feed, too, was of inferior quality and as fewer animals were

[2] A. D. Hall, *A Pilgrimage of British Farming, 1910–1912* (London, 1913), pp. 334–5.
[3] This is noted by John Gibson who was the editor of the *Cambrian News* (Aberystwyth), at the time; J. Gibson, *Agriculture in Wales* (London, 1879), pp. 26–7.
[4] Unpublished manuscript in private possession.
[5] J. H. Clapham, *An Economic History of Modern Britain*, vol. i (Cambridge, 1926), p. 456.

then kept per acre than now, the manure was commonly inadequate in quantity and quality. Hence arose the problem of so applying the manure in co-ordination with the crop rotation that all the cropped land would be manured regularly while in any particular year fresh manure would be provided for the crops that needed it most. This was done by applying most of the manure to the 'potato field', in which many other crops were grown. Annually around 1900 a 260-acre farm which kept horses for haulage work in addition to the usual farm stock, applied the manure of twelve horses and over a hundred cattle and pigs to the 'potato field' of fourteen acres, a different field being used each year.[6] The 'potato field' was one of the key linkages between the agricultural and social systems, as will be shown below.

II

Just as different systems of land tenure can be found in different areas of Europe, so can land be parcelled out into farms in different ways. At the turn of the century much of the downland farming area of southern England was divided into large farms ranging from four hundred acres to over a thousand acres, each capable of independent operation and able to provide regular employment for virtually all the staff that was needed. There were other areas in which the land was divided into small units, while in the Isle of Axholme (Lincs.) land was still parcelled into strips and 'a comparatively large holder' had no more than some forty acres which were held in up to a hundred strips.[7] The agricultural system varied as between such regions, and with it, the social system.

In South Cardiganshire the land was parcelled into units on a continuous scale ranging from the gardens and single fields attached to cottages, at one end of the scale, to 'large farms' of over 150 acres at the other. There was no specialization of product or technique according to the size of the holding, for on each there was produced all that could be of the range of staples locally grown for man and beast. The smallest holding that could provide the whole range of local products was a farm of thirty to thirty-five acres, and of the 203 occupied houses in Troedyraur parish in 1901, 157 had less than thirty acres of land. Thus granted that measure of a subsistence

---

[6] Unpublished manuscript in private possession.
[7] A. D. Hall, pp. 101–7.

element in the economy that has been indicated, 157 households were in some measure dependent on the other forty-six.

Thirty to thirty-five acres was the smallest unit that constituted a 'farm', capable of supporting a man and his wife without a regular source of supplementary income. Such a farm had its own equipment so that it need not borrow from anyone else, and as it was worked with one pair of horses it was known as 'a one-pair place' (*lle par o geffyle*). Smaller holdings could not support the occupants unless they earned an additional income. It was practicable to keep a work horse on a holding of fifteen acres and common on more than twenty acres. These were the 'one-horse places' (*lle ceffyl*) and their occupiers were in fact the auxiliaries in the agricultural system, slaughtering pigs, dealing in butter and eggs, and hiring themselves out as hauliers. This last employment was particularly important before the development of motor transport, for all goods for shops and private houses had to be carted from the nearest railway station and a haulier might be four days a week on the road and the rest of the time on his holdings. Holdings of twenty to thirty acres possessed the necessary equipment to till their own land, but for reasons to be discussed later (p. 47) both they and many of the smaller farms were dependent on larger farms. As holdings smaller than twenty acres lacked the full complement of farm equipment, they were necessarily connected with particular farms in order to cultivate their land, and owed obligations to those farms for services performed for them. This is illustrated by the following example. A holding of nineteen acres kept a maximum of six cows, and had two acres of land under corn. It also cultivated ten rows of potatoes and four of mangels annually. A neighbouring farm sent two men with a pair of horses apiece to till the two acres, which they did in one day. A cart was sent to carry manure from the nineteen-acre farm's yard to the plot where the potatoes and mangels were to be planted. At the hay harvest the larger farm again sent a cart to help carry the hay. There was no cash payment for these services, but a labour debt was due at harvest time. As in this case, so in many others, did particular farms and particular 'horse places' remain linked together year after year.

Holdings that were yet smaller (under fifteen acres) did not keep horses but kept cows to provide 'milk and butter sustenance'. They were classified in the local terminology into 'a cow place' (*lle buwch*), a 'two- (or three-) cow place' according to the number of cows that their land could support. On such a holding oats and potatoes were grown for feeding-stuffs, but as there was neither horse nor plough the occupier was dependent on a farm to perform the necessary work. In many cases these holdings kept their corn in a farm's stack yard if there was any particular difficulty about storing it at home. It could then be threshed when the farm's corn was threshed. To avoid hiring the threshing machine separately in order to thresh their own small quantities of corn other holdings carried their corn to a neighbouring farm in order that it could be threshed along with the farm's corn. To these ends each occupier of a 'cow place' was associated year after year (and his family in many cases for generation after generation) with a particular farm, and he and his family were under obligation to that particular farm for the work undertaken by the farm. Should the farm fail to perform the work expected of it, the occupier of a 'cow place' had no option but to make the best of a bad job. In one instance, around 1900, when a farm failed to plough for a man who held 12½ acres he had nothing to do but spade-dig 2½ acres of grassland with the help of his family. Such holdings were occupied by people in full-time employment as craftsmen and labourers so that digging the 2½ acres had to be accomplished in 'leisure' time.

There were sixty-six of these smaller holdings (under fifteen acres) in Troedyraur parish in 1901. Thirteen were occupied by widows and spinsters, and they (with others who occupied cottages without land) undertook the incidental tasks of the countryside. They knitted stockings, they were specialist makers of oatcakes, and visited farms to work in that capacity. They collected stones from the hayfields and placed them in roadside heaps to earn payment from the highways authority; some worked as quiltmakers, as dressmakers, and others specialized in the trade of making women's garters. They were casual workers at farms where they were paid for an occasional day's washing or potato-sorting. Each one was connected with a particular farm in relations of interdependence as is explained later.

Thirty-six of the eighty holdings of less than thirty acres were parcels of farms, held directly from those farms whether the farmers were freeholders or tenants. Each occupier was known as the *deilad* of the relevant farm and was referred to as *deilad* with the farm's name appended, e.g. *deilad* Pantrodyn. The dictionary meaning, if the phrase may be used, of *deilad* (pl. colloquially *deiladon*) is a tenant, that is any tenant. But the word was not used in that sense in

local parlance. Instead it bore two meanings. It referred to the
tenant of such a parcel as has been noted above but not to the tenant
farmer who held from a landlord. It also referred to someone who
was under obligation to another, who owed allegiance to another,
who was a retainer of sorts. In this sense the word was not uncommon in religious language where the Christian was referred to as
deilad Crist, the person who owes allegiance to, is under obligation
to, and is a follower of Christ. Someone holding a parcel of land
from a farmer was a *deilad* in both senses.

Labourers were engaged by the day or the week, not by the year
(as servants were) and when a labourer occupied a holding as a
deilad he was not necessarily bound to the farm in respect of his
normal work. That varied as the farm's need for outside labour
varied over the years (p. 106 below). But he was obliged to work at
the farm, for payment, when required to do so. This can be seen
in the details of the following agreement between the tenant of an
eight-acre holding and his landlord (who was not a member of the
gentry):

Agreed with Daniel Williams of Bwlch Blaenafon farm etc. For
£5 5s. per year, and he is work for 6d. per Day all round the year 2
meals of meat from Michas [Michaelmas] to St. David's Day and from
there 3 meals. Do. to Michas and he is . . . bound work with me whenever I will and not to go anywhere else without my consent and he is
to pay the above rent, taxes and Tythes belonging to the above mentioned farm, he is now in possession at the 26th day March 1849.

David Davies.[8]

In other words the tenant was not free to dispose of his labour
according to his own wishes. This is noted in connection with a
matter to be discussed below.

Now that the smaller holdings have been considered and their
place in the agricultural system noted and now that it has been seen
that they were not independent self-contained units in that system,
it is possible to discuss the farms on which the smaller holdings were
dependent. Though they secured their cash returns from the sale of
animals and dairy produce, yet it was crop cultivation with its
seasonal labour demands that controlled much of the farm's
activities. It is now difficult to appreciate just how much time was
spent following the plough and pulling harrows and rollers when the
motive power was provided by horses. One man was continuously
employed with each pair of horses for months on end in order

[8] N.L.W., MS. 9956B.

to prepare the ground and complete the sowing in due time. The traditional date by which cereal sowing should have been finished was the last Friday in April at Newcastle Emlyn and the following day, Barley Saturday, at Cardigan, two of the market towns of South Cardiganshire.

A man who farmed a small farm could himself manage to plough and harrow his land with a pair of horses so as to complete the sowing in time and have sufficient oportunity to see to the rest of his work, and in particular, at that time of the year, to the marketing of stock. As the only feeding-stuffs available were those raised on the farm, it was possible to carry more stock in summer than in winter, so that animals were sold off from the farms during September and October to reduce their number to that which the farm could support during the winter. Correspondingly during the spring, stock was bought to raise the number to the maximum that the farm could support during the summer. The fairs at which animals were purchased started in Newcastle Emlyn in March and continued during April until the most important one, May Fair, again at Newcastle Emlyn. These fairs therefore coincided with the season for preparing and sowing the land, and only on a small farm could the farmer himself attend to both duties.

Thus on larger farms a man was needed to work each pair of horses, otherwise the land could not be prepared in time to complete the sowing in season. (All work with horses, including ploughing, was the work of a farm servant (*gwas*) as opposed to a farm labourer (*gweithwr*), who was occupied with such tasks as hedging, ditching, and the like. The distinction between *gwas* and *gweithwr* is treated fully on p. 77). This establishes a relationship between farm work and the farm's staff, for one servant was needed per pair of horses. Hence the expressions 'a one-pair place', a 'two- (or three-) pair place', etc., signified not only the number of work horses kept but the size of the farm's staff.

A 'considerable place' (*lle jogel*) of one hundred acres had some twenty acres under corn annually. A servant working a pair of horses could prepare this ground in time for sowing providing he started early in January and was then engaged for four months on end in ploughing and harrowing. But difficult weather conditions, whether hard frost or prolonged rain, could delay his work, and a 'considerable place' would frequently have two pairs of horses. As it also supported more milking cows, pigs, and poultry than a smaller farm it needed more hands than one extra servant to work it, and its staff consisted of a farmer and his wife, first servant (*gwas mowr*), second servant, a maid, and a 'boy' or 'petty servant' (*crwt, gwas bach*).

Few farms kept more than two pairs of regular work-horses, but 'considerable places' and 'large places' (of over 150 acres) kept breeding-mares and sold their progeny as four-year-olds. They could thus put additional pairs of horses to work if that was desirable and the staff available. There follows the example of the staff of a 'large place' of 220 acres. The farmer did virtually no manual work, being fully employed in management and marketing. Some forty acres were under corn each year, and the farm was a 'two-pair place' so that there were two servants, but as the farm reared horses to sell as four-year-olds an extra pair was put to work if need be, worked by the farmer's son. A labourer saw to the work which did not require horses, while there was a 'boy' at every one's beck and call. The farmer's wife headed the female staff, which consisted of herself and two maids.

It has been noted that the smaller holdings (of less than thirty acres) were dependent on farms as they lacked either the equipment or horses (or both) necessary to work their land. But the smaller farms were in one respect as dependent upon larger farms as were the cow places and one-horse places. Except for a few farms along the coast where it was possible to grow seed corn, farms received their cash income mainly from the sale of cattle and casked butter. Many of the calves reared on the farm were sold off as store cattle in due course as the land could not carry them all, while others were kept as milk cows and eventually sold off as fat cattle. Even the smallest holdings kept cows, for these provided 'milk and butter sustenance', while cottagers (without land) had to content themselves with buttermilk. The cow places could not keep the progeny when the cows calved as they lacked the necessary ground and feed, so the calves were sold to dealers and to butchers to provide veal. Consequently all holdings of all sizes from a cow place to a 'large place' (*lle mawr*) needed the services of a bull. But bulls, being expensive animals to rear and feed, were rarely kept on farms of less than sixty to seventy acres, never on those of less than forty acres. Thus while all of the 133 holdings were dependent on bulls, only twenty-three farms kept them: that is, five holdings out of every six were dependent on the sixth, which was a farm of at least sixty to seventy acres, and kept a bull. Cows were generally taken to the bulls during June, and calved during the spring; the calves then fed on summer grass supplemented with skim milk. Consequently there was an annual 'procession' each summer, when people led their cows to the yards of those farms which kept bulls. If, as was frequently the case, there was no man on the farmyard, the maid released the bull. For this she was given a penny; as was explained to me, 'It was not in payment: it was the custom'.

As has been observed, in 1901 there were 203 occupied houses in the parish of Troedyraur, two of them being mansions. There were forty-six farms (that is, holdings large enough to support man and wife in full-time occupation); twenty-one 'one-horse places' (*lle ceffyl*) where the occupiers were partly dependent on additional earnings; sixty-six 'cow places' (*lle buwch*) where the occupants had other full-time occupations, and seventy cottages with no more land than a garden. It is to the cottagers that we now turn. Their position can best be explained by reverting to certain of the considerations that have already been mentioned.

It has been explained above how five holdings out of every six were dependent on the sixth, which was a farm that kept a bull. This was one reason why no farm of less than sixty to seventy acres could be worked as an entirely independent unit. There was another reason, applicable not only to 'one-pair places' but to all farms, perhaps most particularly to 'large places'. When anyone brought or sent a cow from any holding to another for a bull's service, payment was not in cash but in labour. It was understood either that one day's work at the hay harvest was due for each cow served, or that an occupier who did not keep a bull had all his cows served in return for providing labour throughout the hay harvest. This introduces the second reason why farms could not be run as independent self-contained units, namely, seasonal labour needs.

The position is best explained by taking the example of one particular farm, a 'large place' of 220 acres, at the corn harvest. This was perhaps the most serious period of the year. Unlike their elders, youths regarded the hay harvest as a sportive occasion when young men and women worked in the same fields in a festive mood. But even the youths took the corn harvest seriously. In a short time it was necessary to harvest the barley and oats to provide animal feed, and to garner the wheat 'for the use of the house'. And if the wheat sprouted in the ear during a wet harvest so that it made a sweet and almost unpalatable bread it nevertheless remained for the next twelve months as much an item of diet as if the corn had been harvested in good condition.

A farm of 220 acres had around forty acres under corn annually, the acreage varying slightly with slight differences in the sizes of fields in the proper place in the crop rotation in successive years. The crop was cut with a scythe before the introduction, first of an attachment to the hay-mowing machine that enabled it to cut corn, and then of the reaper and its successor the self-binder. (The reaper cut the corn and laid it in bundles that were convenient for binding into sheaves, but did not itself perform that operation.) The first mowing machine that I have been able to trace was introduced into this area in 1891, and the first self-binder, more commonly known as 'the binder', in 1894. During the corn harvest regular hours were usually worked by the scythesmen, and it was one day's work for one man to cut one acre of standing corn. Thus forty working days were needed to cut the farm's cereal crop, after which the corn lay loose on the ground.

One of the arts of cutting cereals was to lay the corn in a manner convenient for those who would have to bind it into sheaves by hand, and to this end an attachment known as a cradle (*cader*) was placed on the scythe. Oats were cut when they were of 'dove hue' (*lliw'r glomen*), that is, unripe; barley when rather riper. Both were allowed to lie on the ground for a period, ideally of nine days, 'to acquire a tan' which was held to loosen the grain in the ear and thereby facilitate threshing. Wheat was allowed to ripen before it was cut, and then bound 'at heel' (*wrth gwt*, literally, 'at one's tail') by women, one of whom followed each scythesman and used a part of the crop with which to bind the corn into sheaves as she went along. Thus it needed as many working days to bind the wheat as to cut it; barley and oats are considered below. In some areas of mid Cardiganshire farmers continued to reap wheat with a hook for many years after they had started using the scythe to cut barley and oats, possibly for eighty years or so. The immediate reason was that wheat grows taller than barley and oats, so that it did not lie as conveniently in a scythe's cradle. Other reasons are discussed below. What is relevant here is that even more working days would be needed for the harvest at farms where wheat was reaped (*medi*) with with a hook rather than struck (*taro*) with a scythe.

When I inquired whether it took as much time to bind barley and oats as to cut them I received varying answers. Some men with long experience of the work insisted that it took as long to bind corn as to cut it: others equally experienced claimed that corn could be bound in as little as two-thirds of the time needed to cut it. This variation was to be expected for there was one consideration that varied from farm to farm and affected the binder but not the scythesman, namely the presence or absence of thistles in the crop. If there was a good deal of thistle the binder's hands would soon be bleeding, though some people's hands were so hard that thistles could not penetrate them, and the binding took a longer time. If it is assumed that one could bind a given acreage of barley and oats in three-quarters of the time needed to cut it, there will be no serious overestimate of the time needed in order to bind the corn. Hence cutting the corn of a 220-acre farm and binding it into sheaves demanded seventy to seventy-five working days.

This work completed it still remained to carry the sheaves from where they lay on the ground and to place them together in stooks to allow the corn to dry; it was then necessary to carry the stooks together and build them into small stacks (*sopyne llaw*) which kept the crop safe from rain for some time. Later, these were made into larger stacks (*sopyne penlin*), which could not be thatched until a certain quantity of corn was threshed to provide thatching straw, and the required straw ropes made by hand. To complete all the work, close on a hundred working days were needed, and this assumes good weather conditions which did not always obtain.

In the event of wet windy conditions that laid and tangled the corn on the ground it was impossible to use a scythe with cradle attached, so that it was necessary to cut the corn with a 'bare scythe' (*pladur foel*) or a reaping-hook. Apart from the added difficulties of cutting the corn, binding it was then very laborious work indeed. I was told that under these circumstances the reaping-hook had been used in the village of Ffostrasol in 1924, on a farm within three miles of Cardigan in 1946, and as late as 1951 on a farm adjoining the main coast road from Cardigan to Aberystwyth.

When there was prolonged dry weather during the growing season the growth was so short that it was necessary to 'hand reap' (*dwrn fedi*), seizing the whole crop handful by handful while reaping it with the hook. In such years the straw was too short for the use of the binders, and it was then necessary to make enough straw rope (*rheffyn pen bys*) by hand in the stack yard out of the previous harvest's straw, in order to bind the crop into sheaves. This was considered the mark of a difficult harvest. In some dry seasons it might be necessary to treat the whole crop in this way; more commonly only corn grown on dry south-facing slopes was affected. In the driest years (1887 and 1893 were mentioned to me) there was so little growth that the corn could not be reaped at all, and there was nothing to do but pull up the whole crop by the roots (*llafur cito*).[9] The last instance of this that is known to me was on a dry slope on a mid Cardiganshire farm in 1921. Needless to say, under any of these conditions a hundred working days is a serious under-estimate of the labour needs of a 220-acre farm during the corn harvest.

The farm's staff consisted of eight adults including the farmer (who was fully occupied with marketing and management) and three women, the farm wife and two maids. There were only six working days in the week. At the same time there were many tasks that could not be put aside, such as caring for the cattle, milking, churning, and making butter, feeding the pigs and poultry, collecting eggs, caring for the horses, preparing food, washing and laundering, cheese-making (for cheese was an important staple that people had to prepare while summer milk was available), the general work of the farmhouse, and various other and essential items of work. It is clear that with the changeable weather conditions of south-west Wales, the farm's staff could not safely garner the harvest successfully year after year without outside aid.

In this area and during the period within living memory, farms did not co-operate during the corn harvest: all informants are agreed on this point, including one who started full-time work on a farm in 1883. Detailed diaries starting in 1870 confirm this. It is also consistent with the fact that a farmer of, for example, sixty acres, who did not own a bull paid his labour obligations (whether personally or by sending a proxy) to a larger farm for a bull's services, not at the corn harvest but at the hay harvest. Each farm harvested its own corn independently: the extra labour needed was not provided by co-operation between farms.

Yet the corn harvest was so important that the supply of labour could not be left to chance. It was provided through the practice of

[9] Meteorological records confirm that 1887 was an unusually dry year in south-west Wales, and that the early growing season was exceptionally dry in 1893. See H.M.S.O., *British Rainfall, 1887* (London, 1888), p. 221, and H.M.S.O., *British Rainfall, 1893* (London, 1894): Monthly Rainfall Tables.

'setting out potatoes' (*gosod tato mâs*) whereby cottagers set out rows of potatoes in farmers' fields in return for labour at the corn harvest. This was the 'work debt' (*dyled gwaith*) or 'harvest debt' (*dyled cynhaeaf*) of South Cardiganshire, the 'potato debt' (*dyled tato*) of mid Cardiganshire, and the 'potato duty' (duty *tatw*) of North Cardiganshire. The debt was one day's work per row of potatoes to be paid by the cottager, or, as was frequently the case, by his wife. More accurately the debt was one day's work for the load of manure provided by the farmer for each row of potatoes. Cottagers inquired of one another not 'How many rows (of potatoes) have you?' but 'How many loads of manure have you?', and it was the practice to speak not of individuals setting out potatoes but of families.

The number of rows set out by a cottager family varied according to the size of the family, according to the size of the garden, and according to the number of pigs kept in the sty at the bottom of the garden, for they were largely fed on potatoes. Two pigs were kept if possible, one 'for the use of the house' and the other to be sold to pay the rent. A family might have as few as two rows of potatoes 'set out' or as many as eight, usually all at one farm but sometimes divided between two farms. At the turn of the century many South Cardiganshire men worked in the coalfields of Carmarthenshire and Glamorgan while maintaining their homes and families in Cardiganshire. They took their holidays in August and September to return home to pay their families' work debts on the harvest fields. Around Llandysul a concentration of cottager railway workers who set out potatoes allowed some farms to reap wheat with the hook as late as 1910 to 1912. Similarly all dwellers in a hamlet set in one of the steep-sided narrow valleys that cut through mid Cardiganshire to the river Teifi set out their potatoes in the fields of the one large farm that had suitable ground, which enabled this farm to reap its wheat with a hook as late as 1912. The labour needs of a 220-acre farm during its corn harvest have been discussed above. About a hundred working days were needed to harvest the crop. Twenty-four cottager families set out their potatoes at the farm, with an average of four rows each. This provided ninety-six working days for the harvest.

Some time in March the farmer opened the potato rows (which were of a standard length of 100 to 120 yards) with the moulder plough and at either end of each row were placed stakes bearing the cottager's initials. Some owned branding irons for this purpose. The farmer carried a cartload of manure to each row, and informed those cottagers who set out potatoes at his farm of the day on which potatoes would be planted. They were expected to come regardless of other employment in order to plant the farm's potatoes first and then their own. The farmer closed the rows and usually cleaned the crop of weeds in due course. The cottagers' wives were expected to help at the hay harvest but this did not contribute towards the 'work debt'; instead they received quantities of butter known as 'debt butter' (*menyn dyled*); indeed, in days of poverty the meals provided for the wives and children were a sufficient inducement. At the corn harvest the 'work debt' was due, and when the potato harvest came the same people, working with hoes, harvested the farm's potatoes and then their own. That evening the farm's carts delivered each cottager's load to his door.

Even within such a small area as South Cardiganshire there were variations in the details but the general principles were constant. The 'work debt' was for the manure provided by the farmer, so that if a cottager could provide it he owed no 'work debt'. Where the central ridge of South Cardiganshire rises to moorland, turves were burnt and the ash was kept by the cottagers to be collected late in June by the farmer, who used it as a fertilizer before sowing swedes. The ash of culm fires was available along the coast, and cottagers everywhere kept pig manure to be carted to the farms. In mid Cardiganshire it was customary to differentiate between 'manure potatoes' (*tatw dom*) and 'measures potatoes' (*tatw mesure*): the former were potatoes sown in rows for which the cottager had provided manure and were therefore free of 'work debt', while work was owed for the latter. In some localities the 'measure' referred to the practice by which work days were owed not according to the number of rows set, but according to the amount of seed potatoes planted. Around Llanybydder four days' work was owing per Winchester bushel of seed potatoes.

When the cottager provided the manure the farmer's return for the 'manure potatoes' was help to set and harvest his own potatoes; he received 'work debt' as well for the cottager's other rows, the 'measures potatoes'. In some South Cardiganshire localities the farmer weeded the cottager's crop only if he provided an additional day's work at the hay harvest, and the debt for a row of potatoes was a day cutting corn but a day and a half of binding corn.

Reference has been made above to very small holdings which were parcels of farms, and occupied by *deiladon* (see p. 44). One farm of 320 acres contained six holdings of five acres apiece. Two cows and two pigs were kept on each of them, and each occupier set out five rows of potatoes, providing thirty days of 'work debt'. (There were others as well who set their potatoes at this farm.) On this particular farm each of the *deiladon* was at harvest time given a small stack of corn known as *sopyn deilad* ('the "retainer's" stack'), that he himself had made and which contained about forty-eight sheaves. The occupiers of these holdings were day labourers and were bound to work at the farm (for payment) at any time of the year if required; their wives were obliged to help at the hay harvest. Seven cottages stood on the land of an eighty-acre farm, and their occupiers (with others) set out potatoes at the farm. Little if any profit derived from the rent of these cottages, but they provided an assured labour force. The farm's pair of horses would plough the plot of land that an 8- to 10-acre parcel had under corn, in one day, and if the holding lacked a barn, the harvest would be stored on the farm's premises, in return for harvest labour. When mowing machines became common the farm also mowed the parcel's hay while on very small freeholdings the occupier, who being a free-holder was not a *deilad*, was at a disadvantage in that he had to mow his own hay with a scythe, frequently after dark, having worked elsewhere during the day. Interdependence and reciprocity involved farmers, occupiers of 'cow places', and cottagers alike.

The practice of setting out potatoes can properly be said to con-stitute a linkage between the social and the agricultural systems. On the one hand it connected particular cottagers into a specific group connected with a particular farm and its staff, there being such a group connected with every farm. On the other hand it provided a method of systematically manuring the land as has been described above, and it provided too for the additional labour needs of each farm on certain occasions in the course of the year. The potato-setting group constituted the farm's *medel*, which may be rendered 'work group'.

The word *medel* derives from *medi*, 'to reap', and while '*medel*' was strictly a group of reapers the word was used in a more general sense to designate the work group whether its task was reaping or some other occupation. Thus at the hay harvest one would not ask 'How many people were there?' but 'How big was the *medel*?'

People spoke of 'the holding's *medel*' (*medel y tyddyn*) and the
common expression for a group engaged in binding (not reaping)
corn was 'the binding *medel*' (*medel rwymo*). Herewith specific
instances of work groups: 25 cottager families constituted the work
group of a 260-acre farm; 24 families of a 220-acre farm; 9 families
of a 80-acre farm; 8 families of a 78-acre farm; 6 families of a 63-acre
farm; 5 of a 57-acre farm, 3 of a 34-acre farm. (As soil, aspect, slope,
and other factors affecting fertility and suitability for cereal cultivation varied from farm to farm we should not seek a consistently
precise ratio between the size of the farm and the number of its
work group. Further, no account has been taken here of occupiers
of cow places and one-horse places that owed work debts.) Everyone was familiar with farm work. Farmers and cottagers were in
personal contact, they were familiar with one another's affairs, and
they knew of each other's family relationships and histories. While
'setting out potatoes' linked particular individuals into concrete
groups, it also linked two major social categories in that it was
essential to both, a social institution directly related to the needs
of working the land. When those old enough to remember were
asked what machine did most to change rural life no one mentioned
the tractor, all mentioned the corn binder which rendered the
harvest work debt unnecessary. It is true that the question concerned
a matter of opinion rather than a point of fact; it is worth noting
that the opinion was unanimous. It can fairly be said that potato-
setting groups were functional groups such that in their absence the
society would have been different in other respects besides.

The work group was, as has been noted, the connecting link between farmers and cottagers, and perhaps above all else it was the
feature of the life of that time which caught people's imagination.
Co-operative work at the harvest provided occasions for good
fellowship, a sense of shared endeavour and pride in attainment, and
a feeling of companionship and personal 'nearness' which gave a
deep satisfaction to those who worked together. This is reflected in
the sentiment to be seen in the quotation that follows below. As has
been stated elsewhere in this study South Cardiganshire preserves
a popular literary tradition of some antiquity. When one poet who
had himself worked as a member of such a group published a
selection of his work he gave pride of place to a poem which he ends
by saying that should he ever reach heaven, as he surely shall
because his friends are there, it will be an occasion when the

members of the work group will meet again. He lists them and closes with the words

> Y darlun eto'n gyfan
> A'r fedel yno i gyd

—the picture again complete and the members of the *medel* present one and all.[10]

The mutual obligations between farmers and cottagers were operative throughout the year. The children of cottagers spent much of their leisure time in and around the yards of those farms where their parents set out potatoes, doing odd jobs, getting in the way, and running messages. Some farms gave to the child who had been most helpful during the year the first choice of piglet when the cottagers purchased the pigs that they themselves would rear. (They did not necessarily buy at the farms where they set out potatoes for all farms did not keep breeding-sows.) Cottagers' wives called at the farm for buttermilk once or twice a week, whenever the farm had been making butter. One informant recalled that she remembered, as a young girl around 1900, cottager after cottager bringing her bottle to fill, and between them almost emptying the tub of buttermilk that had been put out for them. Seeing her concern her mother turned to her with the words 'Don't worry, harvest time will come'. A comparable reciprocity is to be seen in the following incident. One man recalled how as a boy he went with other children to a farm where their parents set out potatoes but where his parents did not. The children were offered food at the farm and he accepted along with the rest. The farmer's daughter then realized that he was present and the boy was stricken with shame when she said to her mother, 'One of Martha David's children is here'. When he told his mother of this she replied, 'They owe us no debt'.

During the summer when cheese, one of the staple foods, was made, the cottager's wife called at the farm to obtain curds and whey, and similarly when oats had been ground she received oatmeal from the farmer's wife on calling at the farm. The cottager received straw (to use as pig's bedding) at the farm, or failing that collected rushes from the farm's land as a substitute. He received straw too when it was needed for the household's mattresses. At Christmas time he received a measure of swedes, and a quantity of meat if and when a bullock was slaughtered for the use of the

---

[10] D. (Isfoel) Jones, *Cerddi Isfoel* (Aberystwyth, 1958), p. 15. See also A. Jones, 'Y Cynhaeaf', 'Medi', *Cerddi Alun Cilie* (Swansea, 1964).

farm. When the hedgerows, with the bushes and trees that normally grow from them in this area, were cut back during the winter, it was the normal practice for the farmer to take what branches he could lift with a pitchfork and to leave the rest to be gathered by the members of the farm's work group for their own use. The cottagers picked wild fruit, ferns (for pigs' bedding), and twigs at those farms where they set out their potatoes and paid their work debts. The wild products of the countryside were allocated on the basis of potato-setting groups. The arrangements indicate too that cottagers were in part sustained by what the farmers left.

(Earlier in the nineteenth century they had received more of their earnings in this way: before winnowing machines were introduced cottagers, who helped to winnow by tossing the threshed grain in the open air, received payment in the form of the inferior grain (*gwanyd*) sorted out in the winnowing process. Before extra machinery was installed in corn mills and kilns, they received a measure of corn for sieving out corn from mill dust. As long as combs were used to prepare wool for spinning, they were fed 'as one of the family' and received food to take home with them for doing the work.)[11]

Throughout the hay and corn harvests, cottagers' wives and children received food at the farm, while labourers regularly employed at the farm received a measure of milk and a loaf of bread each Saturday night. During the harvests each of the cottagers' wives received a 'home supper' (*swper adre*), that is food to take home with her, as well as the food she had consumed at the farm at supper time. The 'home supper' usually consisted of a quantity of flour measured out to each cottage woman from a bowl kept for that purpose, according to the number of work days she had rendered. She might choose a loaf instead, but in either case a wedge of cheese was also given. Up till 1870 or thereabouts the farm provided the cottager with a New Year's Day's dinner. More recently the farm ran an annual trip to the seaside in its carts, during the interval between the hay and corn harvests. That day the farm provided the food, but the practice varied whether the outing was solely for the farm's staff, for the children and aged cottagers as well, or for all the potato-setting group. Other farms farther from the seaside observed the alternative practice of providing a picnic for the work group in one of its fields.

[11] R. Jones, *Crwth Dyffryn Clettwr* (Carmarthen, 1848), pp. 46–7.

There were further links between a cottager and the farm whereat he set out potatoes. When the cottager had need of a horse and cart as to carry a load of coal from the railway station, the work was done with the farm's horse and cart, for the payment of an extra day's labour on each occasion. Meyrick, writing of the 'Lower District' (i.e. the south) of Cardiganshire in 1810 noted that labourers required carts to carry turves to their homes, and that this was 'generally paid for by labour in harvest, and at the distance of five or six miles four days' work is exacted for the carriage of each load, and each family requires six loads'.[12] On one farm in 1900 cottagers were required to provide half a day's work apiece at the corn harvest for drawing water from a well on the farm's land. Some farms which included lengths of the Teifi's river banks asked a nominal payment of half a day's work at the hay harvest from cottagers in return for permission to fish their stretches of water. Thus there were various and continuing contacts between cottagers and farms, some of which added to the cottagers' labour debts.

In the discussion above of how a farm's supplementary labour needs were supplied it was found that farms had two sources of additional labour, namely that provided by cottagers and occupiers of cow places during the corn harvest, and that provided at the hay harvest by holders of cow places, one-horse places and the smaller farms in return for a bull's services. That the farms were primarily concerned with a supply of labour is confirmed by the fact that many farms did not keep animals whose services were paid for not in labour but in cash, namely boars. A boar's services would be needed only by those farms (generally of fifty acres and upwards) which kept breeding-sows and it was at these farms that cottagers purchased their piglets. Farms of 224 acres, 195 acres, 144 acres, did not keep boars, while a farm of 182 acres and one of 79 acres did. They were generally kept by corn mills which frequently possessed under ten acres of land but had a supply of corn on which to feed the animals. Farms kept them 'according to the farmer's fancy' (as I was told) but not systematically, for no work debt was gained by keeping them nor was it incurred by not keeping them. This is consistent with the arrangements that have been described above.

Further, the institution of potato-setting has been examined primarily as part of a system whereby a farm's labour needs were

[12] S. R. Meyrick, *The History and Antiquities of the County of Cardigan* (Brecon, 1907), p. 114.

supplied. After recapitulating some of the factors involved this institution can be re-examined as part of the cottagers' economy. If the noble animal of the squire and the farmer was the horse, that of the cottager was the pig. It was pig-keeping, which had an important role in the cottage economy, that made the cottagers so dependent upon potatoes, for pigs were largely fed on potatoes until it was time to fatten them for slaughter or sale. Potato-setting and pig-keeping were closely related institutions, they were very nearly two sides of the same coin.

Farms, as has already been noted, made virtually no use of artificial feeding-stuffs or artificial fertilizers. Their supply of farm-yard manure tended to be insufficient in quantity and inferior in quality. Farmers provided for manuring the cultivated ground by concentrating on the potato field, which was heavily manured. As it was in rotation, each field was so treated periodically. The practice of 'setting out potatoes' gained for the farmers the supplementary labour that was needed and also the manure that was provided for the potato field by the cottagers in payment of part of their work debt. Cottagers gained the many varied items we have noted, and potatoes for their pigs.

A cottager very rarely had fresh meat except when he had killed a pig, and he then gave small quantities of fresh meat to neighbouring cottagers (and received similar quantities when his neighbours killed their pigs). But more important, salt bacon was the basis of the broth that was the main meal of the day for cottagers (and most others) during the greater part of the year. The potato harvest coincided with the beginning of the pig-slaughtering season, which was the colder part of the year, from early October to late March. Thus cottagers fattened a baconer to kill so that they could replace it with a porker to rear on the new supply of potatoes. Baconers were kept 'for the use of the house' if possible, being sold only if financial circumstances were pressing, and the sale of a baconer would be interpreted by others in that light. As baconers were sold dead weight, it was necessary to slaughter them whether they were to be kept or sold. Porkers on the other hand were sold live weight, and the cash proceeds of the sale were allocated towards the rent in the normal cottage routine.

Keeping a pig in fact functioned as a savings bank. Moreover pig-keeping was one of the few ways in which a cottager (and especially his wife) could convert his spare-time labour into cash.

In some areas of Wales his creditworthiness at the local shops depended on whether or not he kept a pig (though I have heard no mention of this in South Cardiganshire). Each pig was an object of family interest, and as much a topic of conversation as the weather. Visitors were taken to see the animal, and its condition was a matter of concern for it was in fact an investment which could make considerable financial demands on the cottager when the time came to buy barley meal wherewith to fatten the pig for sale or slaughter. Pig-killing day was literally and metaphorically a big day. It started about five o'clock in the morning when a fire was lit under the cauldron in order to provide boiling water; it demanded a neighbour's help to catch and hang the pig, the services of a slaughterer, and it anticipated what was considered the finest feed of the year.

Metaphorically, 'to kill a pig' meant any sudden accession of wealth or sign of spending. When someone appeared wearing new clothes he would be asked 'What's happened, have you killed the pig?' It was said of someone seen drunk (and who must have spent money in the process), 'He'd certainly killed the pig last night'. When someone displayed a newly purchased item of clothing or equipment, he would half excuse the expenditure by saying, 'Of course we don't kill a pig every day'. The proverb 'The people of the little houses kill a pig but once a year' has already been noted. It indicates that cottagers could rarely afford any extravagance. Potato-setting and pig-keeping constituted a bipolar arrangement involving on the one hand the farmer's needs, and the cottager's on the other.

It has been noted earlier that there was in this area a series of landholding units ranging from individual gardens and fields attached to cottages, to the 'large farms' of 150 acres and upwards. It has been seen too that between the occupiers of diverse holdings there were continuous personal relations which were of interdependence though not of equality. The range of holdings constituted a major part of the agricultural system; it also constituted a ladder of advancement, and in terms of social structure it constituted a hierarchy of status in which prestige belonged to the farmer, as various considerations show.

When decisions were made about those farming operations that involved various people (potato-setting, and the hay, corn, and potato harvests) they were made by the farmer. The others could not but assent. The ambition of many labourers (which many of them achieved) and of most farmers' sons was to be a farmer. Whenever farms fell vacant there was strong competition for them as the findings of the Royal Commission on Land in Wales and Monmouthshire (1894–6) amply attest.[13] One of the farmers' complaints repeatedly recorded and substantiated in that report was that tradesmen and craftsmen who had the means bought farms, thus aggravating the strong competition for any farms that were available. No farm in the parish of Troedyraur fell vacant and unlet until the mid 1930s, when a clayey, boggy thirty-five-acre farm was attached to an adjoining farm when it could not be separately let. Farmers had in effect a 'club' of their own in the fairs which they attended. (Farm servants attended only two or three fairs a year, the May, September, and November fairs.) As will be shown, meal-time practices at farm-houses distinguished plainly between farmers and others. The occupation of cottages and holdings (according to their size) involved a series of work debts that were incumbent on all but the larger farms, and these last alone had the full range of work debts due to them. Cottagers owed work debts at the corn harvest; occupiers of cow places owed work debts both at the corn harvest (for 'setting out potatoes') and at the hay harvest (for a bull's services); occupiers of one-horse places and the smaller farms owed work debts at the hay harvest to the larger farms (for a bull's services). From this standpoint one-horse places and the smaller farms were classified together in local terminology as 'small places' or 'petty places' (llefydd bach). Only farms large enough to keep bulls were not thus indebted to others, and they alone had the full range of work debts due to them. It is worth noting that the status occupied by an individual in his structural hierarchy was only one of many statuses that he occupied in various social institutions and does not of itself indicate his standing in the society.

Concerning the structural relationships between farmers and cottagers it should be noted that the latter were at the former's command. Farmers sent word by one of the children or by the petty servant on the various occasions when the cottagers' presence was required and they had to set aside any other business of their own. Many people when asked whether they themselves had set out potatoes were indignant at the suggestion. When I raised the matter

---

[13] See the *Royal Commission on Land in Wales and Monmouthshire; Minutes of Evidence*, vol. iii (which contains the evidence relating to south-west Wales), London, 1895.

in a group containing farmers and (one-time) cottagers there was such a sharp division between them as I did not see on any other matter. At one and the same time cottagers stated readily that the work debt was at a definite, precise, fixed rate, and insisted inconsistently that it was endless. I obtained this reaction throughout Cardiganshire. Cottagers considered that too much work was asked of them for what they received, but considerations of status were also involved. This is illustrated by a particular South Cardiganshire usage of the word *cardod*, literally 'charity', but which in this area also meant a small and unofficial measure of quantity. When farmers were asked for examples of this usage of the word, their replies were always in the same form, 'Oh, give him a *cardod* of potatoes', 'Give him a *cardod* of swedes'. No one gave as an example a measure of any crop but those grown in the potato field. Many stated of cottagers, '*Ar gardod we nw'n byw*', 'They lived on *cardod*', referring to the buttermilk, swedes, oatmeal, cottagers received at farms, and which were a part of their returns for the work debt that they paid. What is indicative is that it was the word *cardod*, 'charity', and no other that came to mean an unofficial small measure of quantity. Diaries show that cottagers spent every other working day in August and September and every twelfth working day of the whole year, upon farms. They were part-time farm labourers.

At the same time, many of the cottagers who accepted their lower status, particularly widows and spinsters, were offended if they were not asked to work at farms, for there the women constituted gossiping groups and received their food. Consequently many farms would ask all those members of their work groups who were thus disposed, even when there was no need for so many of them. 'You had to ask them all or they would be offended', as I was told, for despite status differences interdependence demanded that good relations be maintained and set a limit on extremes of behaviour. It remains to note that it is only one aspect of relationships that has been discussed above, that aspect connected with the structure of the groups that have been considered. Actual personal relationships in their fullness varied with the personality of group members, and from time to time.

Having discussed the relationships between members of potato-setting groups, one aggregate that has not been mentioned as yet should be noted, those (other than gentry) who were not farmers but who did not set out potatoes; or alternatively those who while setting out potatoes did not themselves or their wives pay the work debt but hired a labourer to do so for them. These were very few in number, clergymen and ministers of religion, retired master mariners, the close kin of farmers to whom their relatives carted potatoes from considerable distances, shopkeepers, and most (but not all) schoolmasters. Their local independence is to be viewed against the background that has been described, and in virtue of this independence many fulfilled particular roles in the society and in the process of change that occurred in that society.

In South Cardiganshire there was, as has been noted, a fixed and known labour debt for each row of potatoes set out in a farmer's field. Some cottagers who were also diarists kept an annual 'harvest account' (count *cynhaeaf*) in their diaries. Herewith an example (translated):

HARVEST ACCOUNT 1888

	(days)
Striking (with a scythe) in the Yellow Stubble Field	1
Ditto	1
Liza and I reaping (with a sickle) at Troedyrhiw for half a day	1½
Liza and I reaping wheat	2
Striking on the bank and binding Liza and I	2
David of Bwlchygroes, similarly	1
Striking in the Large Field, and Liza reaping	2
Striking in the Large Field, and Liza reaping	2
Striking till 10 o'clock on the bank, and Liza reaping throughout the day	1¼
Binding at Troedyrhiw, one afternoon, and Liza reaping and binding throughout the day	1½
Liza reaping and binding throughout the day and I during the afternoon	1½
Liza binding in the Large Field, and churning	1
Binding in ——— Field and other places	1
Liza reaping in Penlon Field and binding oats and peas	1
Finished binding Liza one day and I half a day	1½
	Settled[14]

When the harvest was over and a farmer met a member of his farm's work group informally, the cottager would ask, 'The debt's cleared now?' and the farmer would agree if such was the case. He might ask for a day's work at threshing-time if there was debt still

---

[14] Unpublished manuscript in private possession.

owing but this was unusual and there was no regular payment of debt when the threshing was undertaken.

The debt recorded in the 'Harvest Account' given above is of twenty-one and one-quarter work days, but this is not the total time that members of this household spent on the farm. In addition the diarist's wife was present at the hay harvest, and the diarist or his wife (or both) at the potato-setting and the potato harvest. The diarist himself was a stonemason, but as the extract shows he had to work in the harvest field to pay the work debt. The entry 'David of Bwlchygroes, similarly 1 day' means that one day's work debt was paid by a proxy. The extracts which follow are from a weaver's diary for the year 1883. He was a bachelor living with his siblings; they set out potatoes at Esger farm, and he was a consumptive.

March.
Sat. 14   We set potatoes 2 rows 3½ loads of manure 9 rows all told between them and (the rows in) John the blacksmith's field.
June.
Sat. 9   . . . earthing the potatoes.
Fri. 29   Finished earthing the potatoes in the Esger field.
August.
Thurs. 30   I started reaping at Esger during the afternoon, in the field on the shoulder of the dale.
Fri. 31   Reaping during the afternoon in the slope field.
September.
Sat. 1   Reaping before mid day in the field on the shoulder of the dale.
Tues. 4   Reaping during the afternoon at Esger.
Wed. 5   Reaping all day.
Fri. 7   Reaping all day.
Sat. 8   Reaping during the afternoon at Esger.
Tues. 11   Reaping all day at Esger.
Wed. 12   Reaping all day at the same place.
Thurs. 13   Reaping and binding at the same place.
Fri. 14   Reaping; I was stooking during the afternoon, and they binding.
Sat. 15   At the harvest today again failed completely about four o'clock too weak and retching.
Wed. 19   Daniel Llwynreos at Esger reaping for me, for 1s. 9d.
Fri. 21   Esger finished reaping today.
October.
Sat. 13   Binding at Esger from morning till half past three.
Mon. 29   Esger. I was there moving corn from nine till 4 o'clock.
Tues. 30   Esger. I was binding there during the afternoon finished binding.

November.
Sat. 3   We were digging potatoes.
December.
Mon. 3   Esger, making an account with them, both of us free (of debt).[15]

These diary extracts record only the work done by the diarist at the farm, not the total work done by members of his household for his siblings worked there too in order to repay their labour debt. During September the diarist spent all or part of twelve of the twenty-five working days in the month at Esger, and it can be seen that when the sick man could not fulfil his obligations he did not recompense the farmer in cash but paid a proxy to work in his stead. One further matter should be mentioned. As long as a family's labour debt was paid it was immaterial who paid it. It appears that frequently the women of a cottage household managed to pay the family's work debt without requiring the men to set their normal work aside in order to help fulfil the household's labour obligations. The men then only paid as much of the work debt as the women could not manage. With reference to the two diary extracts quoted above, it would appear that as the nineteenth century drew to its close craftsmen were less commonly seen paying labour debts in the harvest fields than they had been earlier in the century.

Though cottagers and farmers generally appear to have been unaware of it, potato-setting arrangements were not simply neighbourly agreements. They constituted legal contracts. During my fieldwork, farmers and cottagers invariably stated that potato-setting involved no formal agreement between them. 'You just asked the farmer, "Can we set out potatoes with you this year again?" and he would reply, "Yes you may". There was nothing formal.' But an issue of the local newspaper dated 14 July 1911 reports a court case in which a farmer sued a cottager who had set out potatoes in his field.[16] Following the usual custom the farmer had opened and closed the potato rows, providing a load of manure for each while the cottager provided and set the seed potatoes. The cottager had not paid his harvest work debt; the farmer thereupon prevented him from taking any potatoes from his rows and claimed damages against him. The judge found for the farmer. The usual and acknowledged details of harvest labour constituted the 'implied

---

[15] Unpublished manuscript in private possession.
[16] *Cardigan and Tivy-Side Advertiser*, 14 July 1911.

conditions' of setting out potatoes though there was no explicit agreement whether written or oral. In other words, the custom of the country had legal force and effect. This allows comparison of the custom of different parts of the country with the knowledge that there was legal recognition and enforcement of the differences that existed in the custom of different areas.

The data that I have been able to collect are not comprehensive nor have they the geographical coverage that would be desirable. They do show that the custom varied with topography but not with topography alone. In parts of East Wales (for instance in East Montgomeryshire) cash payment for rows of potatoes has replaced a labour debt (whose details are unknown to me) of an earlier period.[17] In East Radnorshire cottagers paid cash for setting out potatoes. On the other hand in some of the upland parishes of the county where little corn was grown farmers reaped in rotation. They formed co-operative groups so that all the members of the co-operative group attended at each farm in turn completing the reaping in a single day, help from the cottagers being inessential.[18] In certain of the Glamorgan valleys, notably in the Pontarddulais area, miners set out potatoes in return for help at the hay harvest. There was no precise work debt, and the arrangement was that should the miner (who had provided the seed potatoes and set them) fail to come to the hay harvest the farmer took half the potatoes, leaving the other half to the miner. Among the miner's 'rights' as a member of a potato-setting group was the right to hunt over the farmer's land if the farmer was a freeholder or held the game rights of the farm if he was a tenant.

In the hill country of Merioneth relatively little corn was grown and cottagers who set out potatoes at farms helped in return at the hay harvest. There was no precise amount of work due for the number of potato rows set. Farms with uplands in which there were small areas suitable for improvement used the practice of 'setting out potatoes' in order to improve those areas. During the first year the potatoes were set in the turf, and nothing was required of the cottager but to provide the seed potatoes and harvest the crop, which was carried to his home in the farm's cart. In subsequent years the farmer provided the manure and prepared the potato rows

[17] *Report of the Royal Commission on Labour: The Agricultural Labourer,* vol. ii (Wales), (London, 1893), p. 86.

[18] W. Watkins, 'Llanfihangel Rhydithon Seventy Years Ago', *The Radnorshire Society, Transactions,* vol. iv (1934), p. 26.

in the usual way while the cottager set the potatoes and assisted at the hay harvest. Again there was no precise labour debt, and local parlance knew of no expression corresponding to the 'work debt' (*dyled gwaith*) or 'harvest debt' (*dyled cynheia*) of South Cardiganshire. These small upland areas, thus improved, still provide better grazing than adjoining unimproved areas. An alternative custom known as *tatw cyd*, literally 'co-potatoes', also obtained in Merioneth, whereby cottagers set out rows of potatoes but only harvested every other row, leaving the rest to the farmer. (I have recorded this in the district of Pennal, and there is documentary evidence for it in the parishes of Llanfor and Llanycil in the vicinity of Bala.)[19]

In some upland parts of Carmarthenshire there were very few cottagers, but in the western parts of the county near the river Teifi, and in the northern part of Pembrokeshire, the custom was as that of South Cardiganshire. Within Cardiganshire the same custom prevailed in the northern upland areas of Devil's Bridge and Ystumtuen as in the south. These upland areas were not good cereal country. It was impossible to grow wheat and at harvest time children were not allowed to carry the sheaves of barley, which was the bread corn, to place them in stooks, lest they lose a certain amount of grain by shaking the sheaves so that grains fell out of the ears of corn. On the lower ground near the coast of North Cardiganshire the labour debt was three days' work at the corn harvest for each Winchester bushel of seed potatoes planted. It may be that the similarity with the custom of the much more favourable area of South Cardiganshire resulted from the presence of numerous cottagers in both areas, many of those in the north being lead-miners. Or it may be that the similarity was the outcome of sharing in a common regional custom, that of south-west Wales, which once had numerous characteristic features of its own.[20]

[19] *Royal Commission on Land in Wales and Monmouthshire, Minutes of Evidence*, vol. i (London, 1894), p. 284.

[20] Various attempts have been made to suggest a northern boundary for the ancient division of south-west Wales but this is not of immediate concern here. It is sufficient to note that the details of a fixed and known labour debt on farms near to, and to the north of, Aberystwyth are recorded in the *Commission on the Employment of Children, Young Persons and Women in Agriculture (1867): Third Report of the Commissioners* (London, 1870), pp. 128, 130. Details of the labour debt in a North Pembrokeshire parish (St. Dogmael's) are given on p. 40 of the same report. Details of the labour debt in the district of Llanboidy (in West Carmarthenshire) are given in the *Report of the Royal Commission on Labour: The Agricultural Labourer*, vol. ii (Wales), (London, 1893), p. 68.

This possibility is consistent with the statement made in 1815 in the first comprehensive description of the agriculture of South Wales, where the author noted '. . . a kind of feudal connection, such as that which still subsists between farmers and their labourers, in numerous instances, in the Dimetian counties of Cardigan, Pembroke and Carmarthen, where labourers and their families may be considered as heir-looms or appendages to the farms. . . .'[21]

It is now possible to turn to the social implications of these variations in the custom of the country. The structural relationship between farmers and cottagers was a very important constituent of that system of roles and relationships that constituted the social system. Thus this variation in farmer–cottager relationship must be viewed as implying variations of social structure. This may help to explain the assertions and counter-assertions that have at times been made about the nature of 'Welsh society'. Societies 'are not "things" in any material sense. . . . The concept of society is a relational not a substantial one; the only concrete entities given in the social situation are people. What we indicate when we use the term "society" is that these people are related to one another in various institutionalized way'.[22] It has been shown that the institutionalized ways in which farmers and cottagers stood related to one another varied in different parts of the country.

Having shown how the extra seasonal labour required on farms was provided in South Cardiganshire, it can be seen that the arrangements were very closely related to such factors as the low income per acre yielded by an economy mainly dependent on the sale of store cattle and casked butter, the small size of the holdings, and the necessity of raising corn in an area better suited to grass. But it cannot be concluded that the given conditions inexorably produced the system that has been described. Seasonal labour has been a major farming problem in many areas of Britain. Certain parts of southern England have been fertile enough to allow farms to give regular employment to so many labourers that with rather better harvest weather conditions than in south-west Wales they were able to harvest their corn regularly with virtually no outside labour. On some Suffolk farms the farm's own staff 'took the harvest', that is, contracted to undertake the harvesting. This is described by

[21] W. Davies, *General View of the Agriculture and Domestic Economy of South Wales*, vol. ii (London, 1815), p. 283.
[22] J. Beattie, *Other Cultures* (London, 1964), p. 34.

G. E. Evans.[23] A farm's staff consisted of labourers living out. Their cottages had gardens and each labourer had an allotment of some quarter of an acre which corresponded to the potato rows of South Cardiganshire. On his allotment the cottager grew green vegetables, potatoes, and corn to feed to his poultry and perhaps for his family. Farmers were generally unwilling to lend horses so that allotment holders could plough their land, in which case it was necessary to dig it with a spade. If horses were lent, they were lent only on Good Friday and on condition that they were returned to the stable by 'church time' (eleven o'clock in the morning).

At harvest time one of the regular labourers (known as the Lord of the Harvest) contracted with the farmer on behalf of a group in order to take the harvest. This group consisted of the regular farm labourers (less those retained for essential routine work) and any seasonal workers contacted by the Lord of the Harvest. Each man in the group was responsible for providing a gaveller, that is a woman, who with the other gavellers, would rake the mown barley into gavels or rows, for barley was not bound into sheaves but pitched into wagons in the manner of a hay harvest. Wheat and oats were bound into sheaves by a 'tier-up', who might be a man or a woman and who was assisted by a 'bind puller', that is, someone who pulled a number of stalks from the corn and handed them to the tier-up to use in binding each sheaf. Boys and girls still at school, or newly left school, were included in the contracting group to do this work. During the harvest the farmer receded into the background and the Lord of the Harvest with his contracting group took over. It should be noted that in this arrangement the work group consisted of the regular farm workers, along with others whose obligations were not to the farmer but to the Lord of the Harvest, and who had no necessary connection with the farm except during the harvest period. The regular labourers harvested their own corn from their own allotments independently, while their wives and children were admitted to glean at the farm once the farm's harvest was gathered in. In other parts of East Anglia during the nineteenth-century harvest labour needs were catered for by the labour gang system, whereby a farmer contracted with a labour gang (with which the farm had no contact but at harvest time) to undertake the corn harvest.[24]

[23] G. E. Evans, *Ask the Fellows who Cut the Hay* (London, 1956), pp. 85–117 *passim*.
[24] W. Hasbach, *A History of the English Agricultural Labourer* (London, 1908), pp. 192–203.

Flora Thompson has described a hamlet on the boundary between Oxfordshire and Northamptonshire where all the agricultural labourers who lived in the hamlet were employed at one large farm.[25] Each labourer's cottage had a garden attached where green vegetables and fruit bushes, but not potatoes, were cultivated, and in each garden stood a pig sty. Each cottage also had an allotment which was spade dug by the cottager who grew potatoes on half of his plot and corn on the other half. At harvest time each cottager cut his own corn, threshed it with a flail, and winnowed with a sieve. At the farm regular labourers were supplemented with migratory Irish harvest workers, while the regular labourers' wives bound the cut corn into sheaves. Those villagers who were not members of the farm's regular staff (as for example the stonemason) did not work at the harvest, nor did their wives. Harvesting involved no local families except those whose menfolk were regularly employed at the farm.

In parts of Northumberland extra labour was secured by the bondager system which was in operation into the late nineteenth century.

Each farm is provided with an adequate number of cottages having gardens, and every man who is engaged by the year has one of these cottages: his family commonly find employment, more or less; but one female labourer he is bound to have always in readiness, to answer the master's call, and to work at stipulated wages: to this engagement the name of bondage is given, and such female labourers are called bondagers, or women who work the bondage. Of course, where the hind (as such yearly labourer is called) has no daughter or sister competent to fulfil for him this part of his engagement, he has to hire a woman servant, . . . but this is not very common. . . .[26]

In other words, this system generally involved only those employed on the farm and their closest relatives. What characterized the South Cardiganshire arrangement was that it involved all cottagers (including craftsmen) whether or not they or their relatives were farm labourers, and that it involved them intermittently throughout the year, one year after another.

Potatoes were introduced into Wales during the second half of

[25] F. Thompson, *Lark Rise to Candleford* (Oxford, 1939), esp. chapters III and IV.
[26] *Report of the Special Assistant Poor Law Commissioners on the Employment of Women and Children in Agriculture* (London, 1843), pp. 296–7. See also the *Report of the Committee on the Employment of Children, Young Persons, and Women in Agriculture, First Report* (London, 1869), pp. 52 *et seq.* According to Higgs the practice of 'bondage' survived in some measure until the First World War; J. Higgs, *The Land* (London, 1964), note to illustration no. 208.

the seventeenth century but it was a hundred years later before they became the food of the majority of the population in Cardiganshire.[27] They were one of the chief staples before 1800. Evidence is scanty as to whether or not the farm work group pre-dated the introduction of the potato but what evidence there is suggests that potato-setting practices were a new development of an old institution rather than an entirely new phenomenon. Bron-wydd Manuscript 619 in the National Library of Wales, dated 1736 and entitled 'John Warren Esquire of Trewerns Harvest Account' records 'The Names of the Customers bound for Harvest', and the payments made to them. Few other details are found in this manu-script, but it is clear that the 'customers' were those bound by custom to provide harvest labour. Some of the customers fulfilled their obligations in person as in this entry, 'William Lewis Vagwrlwyd Custom 1 Day'. Others did so by proxy as in the following entry

'David Lloyd Custom for   Evan David of Penycnuck
Thos William
1 Day'.

The need for extra labour at the corn harvest was no new feature of the eighteenth century, and it appears that at one time the reward of those who worked at the harvest was the right of gleaning. The men reaped, their wives and children carried the corn to where the sheaves would be built into stacks and then they gleaned each plot of ground once the sheaves had been cleared. (On certain medieval manors of southern England people were allowed to glean in pro-portion to what they had reaped. On some manors a man received a standard payment in sheaves for his customary ration of work, and for extra work a man 'ought to have one woman gathering spears of grain after the reaping'.)[28] The word *medel* which has been trans-lated as 'work group', derives from *medi*, 'to reap', and this of course long predates the introduction of the potato.[29]

*   *   *

When this society is examined as it was at the end of the nine-teenth century it can be seen that it was ordered to meet the needs of working the land. This was not the only way in which the society

[27] R. N. Salaman, *History and Social Influence of the Potato* (Cambridge, 1949); 'The Potato in Wales', pp. 409–23.
[28] W. O. Ault, 'By-Laws of Gleaning and the Problem of Harvest', *Economic History Review*, vol. xiv (1961), p. 215.
[29] What is here described raises a further problem in the history of agriculture: when was the system devised of manuring the land by concentrating on the potato field?

was ordered, and those relationships which have been described do not constitute the whole, but one important part of the social system. The society has been viewed in relation to the land, as an ecology. What has been arrived at is one field of social relationships, namely that which is based on ecology.

That from one standpoint the society can be viewed as an ecological system is not to say that the components of the system worked with a machine-like precision or efficiency. There is no good reason for thinking that the ratio of cottages, cow places, one-horse places, and farms was an 'ideal' one. It was simply workable. It can be shown that five adjoining one-horse places and cow places were established not to help constitute an ideal ratio between farms, horse places, and cow places, but to meet the accidents of the circumstances of one particular family at one particular time. A free-holder of some thirty acres had two sons and two daughters; he divided his holding into five parts, one for himself and one for each child. Had he fewer children he would have divided his land into fewer parts. (This presupposes a notion to be discussed later, that provision should be made for each child in one way or another.)

Some farms gave potato ground to more cottagers than their labour needs warranted lest some cottagers be left unprovided for. One small farm gave potato ground to ten cottagers, while a farm of twice the size had only seven cottagers setting out potatoes on its land. Another rather remote farm could not get sufficient cottagers to satisfy its labour needs; it had no cow places to attract cottagers and the estate lacked the resources to provide the buildings that cow places would require. A farm of seventy-seven acres had eight cottages on its land while another of over 200 acres had only four. 'Working the system' was a matter of making do. Nor is there any compelling reason for regarding this state of affairs as the deterioration of a state that was at one time ideal, or more nearly so.

III

FARM ORGANIZATION

IN 1861 at least nine households in every ten in Troedyraur parish were connected with farms and farm work in one way or another. As can be seen from the accompanying table (Table 2) five households out of every ten were those of farmers and farm labourers and of those who earned their living in part from their land and in part from labouring work. Another four households in every ten were members of work groups which were connected with farms throughout the year though more particularly at harvest times.

TABLE 2

Heads of households in Troedyraur parish in 1861[1]

Total number of households	233
Spinsters' and widows' households	
(i) Number	64
(ii) Percentage of total households	28%
Farmers' households[2]	
(i) Number	38
(ii) Percentage of total households	16%
Labourers *cum* landholders' households	
(i) Number	18
(ii) Percentage of total households	8%
Labourers' households	
(i) Number	54
(ii) Percentage of total households	23%
Craftsmen's households	
(i) Number	37
(ii) Percentage of total households	16%
Miscellaneous households	
(i) Number	22
(ii) Percentage of total households	9%

No such complete survey of the position *circa* 1900 is now possible. But it is possible to compare the details of the Census

---

[1] Source: Census of 1861, recorders' notebooks, in the Public Record Office.
[2] The heads of four farm householders were widows and these have been included among 'Farmers' Households'.

Returns of 1861 with data contained in the school admission registers of the Troedyraur Board School at Rhydlewis for the last quarter of the nineteenth century. (There was no other school in the locality.) The registers state the occupations of the heads of those households whose children attended the school.

The detail in the school admission registers is obviously inadequate in itself if only because it mentions few widows' or spinsters' households though these were numerous in 1861. But an estimate of the position around 1900 is possible by comparing the school admission registers with the 1861 Census data less the spinsters' and widows' households.

TABLE 3

Heads of households (less those of spinsters and widows)

(*a*) Troedyraur parish in 1861

Total number of households	169	
Farmers' and labourers' households		
(i) Number	110	
(ii) Percentage of total households		67%
Craftmen's households		
(i) Number	37	
(ii) Percentage of total households		22%
		89%

(*b*) Troedyraur Board School admission Registers: Heads of households[3]

Total number of households	195	
Farmers' and labourers' households		
(i) Number	160	
(ii) Percentage of total households		82%
Craftsmen's households:		
(i) Number	23	
(ii) Percentage of total households		12%
		94%

As will be seen, these calculations show that nine households in every ten were connected with farms and farm work in one way or another. Almost the whole population was concerned with the way

[3] N.L.W. MS. Cardiganshire L.E.A. 137a.

in which farm work and duties were organized and allocated between men and women and between the various servants and labourers that constituted a farm's staff. It has been demonstrated how the overall nature of the society was related to the needs of working the land for this was shown in the way in which farmers and cottagers were linked with and related to one another by common membership in potato-setting groups. It is now necessary to inquire how the detailed arrangements of farm organization were related both to the work of the land and to the nature of the society at large.

A basic feature of the customary organization of farm work was its division into men's work and women's work. The men were responsible for the cultivation and the general labouring work of the farm while the women undertook the dairying and the work of the farm-house and the farmyard. Correspondingly it took two people to run a farm, the master and the mistress, who were respectively the heads of each division of the work. At the same time the roles of 'master' and 'mistress' carried particular prestige among people who for the most part were dependent on the land for their living in greater or lesser measure.

These two roles were filled by the husband and the wife respectively where the farm was occupied by a married couple. But where the farm was not occupied by a married couple the roles of master and mistress might be filled in any one of a number of ways. If the head of the household had children who were old enough to fill the relevant role then a son became master in a widow's household and a daughter became mistress in a widower's household. She would be addressed as 'miss' as would a spinster sister in a household in which her bachelor brother was master. Where the head of the household was a bachelor or widower without a sister or a daughter who could fill the role of mistress he employed a housekeeper to play the role in farm organization that would be played by a wife were the farm occupied by a married couple. Colloquially the position and the woman who occupied it were referred to as 'housekeeper'.

On the other hand where the head of the household was a woman without sons or brothers to fill the role of master she employed a man to fill that role. He was referred to as 'bailiff'. But should she marry a servant who thereby became master he was referred to as 'gwrwas', which may be rendered 'husband servant' and is discussed on p. 84. In each case those who were members of the work group as well as those who were employed at the farm addressed the male head of the one division of farm work as '*mishtir*' (master). They addressed the female head of the other division of farm work as '*mishtres*' (mistress) or 'miss' as the case might be. After describing the two main divisions of farm work as between the men and the women for which the master and the mistress were responsible, it will be practicable to examine the position of the staff which each controlled.

The mistress was responsible for the female staff, daughters and maids where there were both. She herself prepared the household's food and undertook the butter-making and the cheese-making, perhaps helped by one of the maids, but the greater part of the maids' time was occupied not in the farm-house but on the farm-yard. It was not as domestic servants that maids were employed. Where there were daughters it was they who worked in the house, and if there were maids as well the latter were engaged in farmyard work. On the other hand maids did not work in the fields except during the harvests when they helped in loading the hay and at the ricks, and in binding the sheaves during the corn harvest. The dairying was the mistress's responsibility, and this was not confined to work in the dairy but included much of the care of the milk cattle. South-west Wales was known as an area where a high proportion of female labour was employed on farms. According to the *Third Report of the Committee on the Employment of Children, etc. in Agriculture* (1870) 'in the ordinary routine of farm work, the women servants tend cattle, clean out the stables, load dung carts, plant and dig potatoes, hoe, take up, top and tail turnips, and in the spring drive the harrow, but they are never trusted with horses and carts on the high road'.[4]

The end of the century saw men undertake much of the heavier work noted in this quotation. Women still did all the milking, which was explained as required by the fact that men's hands were too dirty to milk. The milk cattle were kept in the cow house, which was a stone-built building except where the dwelling-house was mud-walled in which case the cow house was likely to be of the same material. On many farms women still cleaned out the cow stalls at the end of the century but if one of the men had taken over

---

[4] *Third Report of the Commissioners on the Employment of Children, etc., in Agriculture* (London, 1870), p. 52.

this duty it was either the second servant or the 'lad', who was the lowest in the hierarchy of farm workers, not allowed to clean the stables, and at the bidding of everyone except the petty maid (*morwyn fach*). The men prepared the meal for the milk cattle but it was the women who fed them and gave milk to the young calves in the calf box. On the other hand the store cattle were kept in a separate shed and the women were not concerned with them for their responsibility was the dairy and with it much of the care of the milk cattle.

The prestige work, however, which was described as 'the highest work of the farm' (*gwaith ucha'r ffarm*) belonged to the men. Their work fell into two major parts, the work of cultivation which entailed working with horses (which was the 'highest work') and the general labouring necessary to maintain and improve the farm's physical condition. This was a society wherein prestige and power were in the hands of the elder rather than the younger people, and children were expected to defer to their parents until they established their own households. The average age of marriage was late (farmers' sons 32, farmers' daughters 28) and freedom to manage one's own affairs came only with marriage, or the death of one's parents, so that from this standpoint it was the elder and married people who were of higher standing.[5] Yet the 'highest work of the farm' was the work of usually young and unmarried men, the farm servants. A farmer could not take one of his own horses without the permission of the head servant who was responsible for the horses. If he did it was regarded as sufficient reason for the servant to break his contract of employment.

A farm servant (*gwas*) was quite distinct from a farm labourer (*gweithwr*). Farm servants were unmarried and generally young men who were engaged to work with horses and who lived in at the farm while labourers were usually married men who lived in their own houses. The general labouring work of the farm such as hedging, ditching, and drainage was specifically the work of the farm labourer. It was with this work that parents threatened lazy children, 'It's pick and shovel work that faces you.' A farm servant became a farm labourer when he married and set up his own home; henceforth he was engaged on a different contract and ceased working with horses. If a farm servant remained single he continued to be a servant, fulfilling a servant's duties; he did not become a farm

5 Information from the General Register Office, for the parish of Troedyraur.

labourer. But servants or labourers who had become too stiff to follow their usual occupations might become cowmen, for some of the largest farms employed them. The care of cattle, which largely brought in a farm's cash income, was for men the work of lowest standing and accepted only as a last resort when nothing else was available. It was considered degrading for a man to milk, an attitude that scarcely changed until milk production became far more important with the sale of bulk milk after the establishment of the Milk Marketing Board in 1933.

The master did no manual work at all on large farms, even at harvest times. (In parts of Cardiganshire it was said that 'a large place keeps one idle man'.) His duties were the allocation and supervision of work on the one hand, and marketing stock and casked butter on the other. This involved him in negotiations with the dealers who frequented the fairs held at the local market towns until auction marts were established during the First World War. As has been indicated he was 'master' not only to the farm's staff but to those cottagers who set out potatoes at the farm. He decided when the various operations connected with potato cultivation were performed, when those who set out potatoes were required to attend for planting, hoeing, and harvesting. It was to him that members of the farm's work group came to request straw for their pigs' bedding, for permission to cut rushes used for the same purpose, or to collect firewood after hedges had been trimmed during the winter and early spring. When it is recalled that twenty-five households, involving more than a hundred people (including children), set out potatoes at a farm of 260 acres, it is easier to comprehend the standing of a master in a farming area.

The mistress was responsible not only for the maids and their work, but for selling whatever dairy produce was sold apart from the casked butter which was taken to the dealers at the market towns. This included pats of butter as well as any milk that was sold. It should be recalled that many people held cow places that provided them with 'milk and butter sustenance' as it was called and did not purchase milk. In addition the mistress sold the eggs which she usually collected from the nests personally. So clear cut was her right to these that people spoke of farmers stealing the eggs of their own poultry, as will now be explained.

Before the establishment of the Milk Marketing Board in 1933 (which pays farmers monthly for the milk they sell to the Board) a farmer received money only when he made sales to the dealers who paid in cash. If his wife was grasping a farmer might himself have no ready money, in which case he might resort to taking eggs surreptitiously, and to sell them. When this happened people had no doubt at all how things stood at the farm involved and spoke of the husband 'stealing eggs', an action that occasioned many a jest. It was still happening well after the end of the Second World War to my certain knowledge. That people spoke of farmers stealing eggs from their own farms did no more than indicate that everyone was familiar with the common arrangement whereby it was the proceeds of the sale of eggs, milk, and pats of butter that the mistress had to run the house. This money was hers and no concern of the master. It was referred to as 'farmyard money' (*arian y clos*) or as 'egg money' (*arian wye*) in Cardiganshire. A mother would say to a child who came asking for spending money, 'What's the matter with you? I've only got "farmyard money" to run the house'. This practice of allocating exclusively to the mistress the money that was received in return for certain farm products was widespread. In some parts of Wales the money was known as 'basket money' (*arian y fasged*), referring to farm wives' practice of delivering butter and eggs to private houses or taking them to market in plaited baskets covered with white cloths.[6] Emigrants took the practice with them and it has been found in certain areas in the United States.[7] Similarly along the border between Wales and England farm wives had the proceeds of small orchards with which to run the house, while in other border areas the practice whereby the wife retained the 'hen moneys' dated from at least the eighteenth century.[8]

Just as the farmer was 'master' to members of the farm's staff, so was a farm wife (or whoever filled her position) 'mistress' to the same group of people. It was to the mistress that cottagers' wives went, or sent their children to get buttermilk after churning. It was the mistress who gave oatmeal to cottagers' wives after the oats had been ground, and it was she who handed out the loaf that was given to each of the farm labourers every Saturday night. At harvest times she presided over the meals prepared for the work group and doled out measures of flour to those who preferred this to receiving supper at the farm. It was to her too that the women of those

6 A. D. Rees, *Life in a Welsh Countryside* (Cardiff, 1950), p. 63.
7 J. West, *Plainville, U.S.A.* (New York, 1945), p. 47.
8 A. Hughes, *The Diary of a Farmer's Wife 1796–1797* (London, 1964), pp. 20, 133–4.

households which set out potatoes at the farm came during the course of the year for small favours and occasional help.

The difference between a farm servant and a farm labourer has already been noted. In order to elaborate on the significance of the difference between them it is necessary to anticipate a distinction to be made in greater detail in a later chapter. It is the distinction between a farm household considered as a family unit on the one hand and as an organized unit of production on the other. A farm family could not always supply from its own resources the labour needs of the farm that it occupied. Under those circumstances one of the ways of resolving the difficulty was by employing outside workers whenever the children's labour needed supplementing. If there were many children of working age they undertook much of the general labouring work as well as the tasks of cultivation, but no farmer would give a son labouring work and employ a servant for the task of cultivation. The first work of the sons was cultivation, with outside people employed as labourers. If there were too few sons for the work of cultivation it was farm servants (single men living in) who were employed to supplement them, not labourers (who were married men living out).

The son and the servant (not the labourer) were equivalent in the way that the farm's staff was organized as a production unit. When people are asked how many servants a farm of a certain size would have employed they find the question unintelligible and unanswerable for the assumption is always that servants were only employed in default of sons. Hence such questions can only be asked about specific farms, and people answer that there were so many sons at home and that so many servants were employed as the case might be. Sons and servants did the same work, one was needed for each pair of horses. Up to a point they were paid in the same way. Servants were engaged on a twelve-month contract each November. They received no weekly wages but went to the master as the farmer's son did to draw sums of money as they required them, which was usually when important fairs were held. The balance accrued to a servant when his contract ended while the son received his portion on being established on his own farm. Sons and servants shared the same sleeping-quarters and in their leisure time they mingled with one another and did not constitute themselves into two separate groups. And they were both concerned with the animals which were the pride of every farm, the horses.

The chief servant had the care of the stables and the horses and no one (including the farmer and his sons) could take any of them without the chief servant's permission. Each son or servant was responsible for his pair of horses, and he spoke of them as '*my* horses'. A son or a servant grew very attached to 'his' horses, and it was by no means unknown for a servant to break his contract when a farmer sold a horse that was in that servant's care. Those responsible for each pair of horses bought medicines for them out of their own pockets and they purchased horse brasses to adorn the horses particularly on occasions when they were paraded, as when a farm's cart took a party of the farm's workers to the seaside, and at the parades of horses that were held in late April or early May (when the sowing season ended) at the local market towns. The servants kept the horse brasses in their sleeping quarters and when a servant removed to another farm he took the brasses with him. One item that he renewed whenever he changed from one farm to another was his whip, while each November, with each hiring, he provided himself with a new pair of boots and a new pair of trousers, which were to last the year. We shall note the significance of this point later. Bootmakers' and tailors' workshops were piled high with new boots and trousers awaiting collection by the end of October and it was after the servants received the money accruing to them that these craftsmen were usually paid for work done during the year. The same system obtained in shops and small woollen factories. They engaged their men from one November to the next, and set out potatoes at various farms to help feed them, paying a harvest labour debt as individual households did. Many of the small factories were parts of farms, standing on or by the farmyards. One informant recalled being one of fourteen sleeping in a farm's storehouse, three being farm servants and eleven woollen workers.

As long as the horses were required for farm work those farms which kept stallions took them from farm to farm to serve the mares during the summer (for a cash fee) and a 'stallion card' (*carden march*) was left at each farm where a call had been made. The cards bore the stallions' pedigrees and they were pinned up on the stable doors where they were the objects of general interest. Sons and servants took great pride in the appearance of their horses, and they were the people who principally engaged in the annual ploughing competitions. Thus there was a general interest in horses, but an especial interest among sons and servants for the horses were their particular concern.

But while sons and servants were equivalents in the way that the farm's staff was organized for working purposes, the son had a position too in the family that occupied the farm. Servants' holidays were restricted to three fair days a year, the May, September, and November fairs. If a servant changed farms in November he then took a week's holiday, but if he re-engaged at the same farm he had only the afternoon and evening of that day unless he lived at a considerable distance in which case he might have time to visit his home before starting work again. A farmer's son would have opportunities to attend agricultural shows and auctions, but one auction a year was the most a servant could hope for, and time to go to the Sunday School's annual festival. Indeed the common complaint at the end of the nineteenth century was that the obverse of what has already been described no longer obtained, namely that while sons and servants were equivalents in the way in which the farm staff was organized as a working unit, the measure to which the servant had been assimilated to the family was less than it had been earlier in the century.[9] (In some parts of Wales farm servants addressed the farmer and his wife as 'uncle' and 'aunt' in mid century but as 'master' and 'mistress' by the end of the century.[10]

At the same time there was one respect in which the position of a servant in a farm household was very intimate and to a degree ambivalent. The young children of the family doted on the servant, and the servant treated the children very affectionately. Not at all infrequently the children were fonder of the servant than of their father and their adult brothers. The servant brought them little gifts each time he visited fairs, he fondled the children, and stood to them in the relationship commonly associated with a friendly uncle able to indulge the children without having to discipline them. This relationship was widely recognized and it was common to hear comments on how fond particular servants were of the children at a farm. One may properly say that it was one of the servant's roles to stand thus to the household's children. It has previously been mentioned that if a farm servant did not marry he remained a servant (rather than become a labourer) and some such servants

    [9] *Royal Commission on Land in Wales and Monmouthshire, Minutes of Evidence,* vol. iii (London, 1895), p. 65.
    [10] H. Evans, *Cwm Eithin* (Liverpool, 1950), p. 38.

stayed with the same family for very many years, seeing the birth and growth of the children, the deaths of some children, and the marriages of others. On occasion they fussed as much at the weddings as the parents did, and might be given places of honour at the wedding festivities. While in the nature of the case they did not become members of the family, nevertheless the attachment between them might be very close.

As distinct from the 'highest work of the farm', the general labouring was the work of a farm labourer (*gweithwr*). There was more than one kind of labourer and the reference here is to labourers regularly employed at particular farms. They were engaged by the year, but unlike servants they were paid by the week, receiving milk and a loaf of bread each week in addition to their wages. They did not work on Sundays but not uncommonly they had their Sunday dinners at the farms where they were employed. A farm's labourer lived in a cow place that was a parcel of the farm, that is he was a 'retainer' (*deilad*, see p. 44). Such a man was known colloquially as *gweithwr* (labourer) with the name of the farm appended, or as *deilad* (retainer) with the name of the farm appended, for example, *gweithwr Gorslwyd* or *deilad Y Figin*. A servant on the other hand was known as *gwas* with the name appended of the farm where he was employed, for example, *gwas Y Wern*. In addition there were other labourers known as *dynion hur* ('hired men') who were casual labourers employed and paid by the week or the day. These were required for the harvests, thatching the ricks, making potato clamps, and other skilled work. They were usually highly skilled men and were acknowledged to be such but they stood at the bottom of the hierarchy of farm workers. Against the background that has been presented the prestige of different farm occupations is easily understood. 'The highest work', namely work with the horses, was that of the family in which the servant shared because he was the equivalent of a son in the farm's work organization. Next came the family's retainers, the regular labourers who were known by the name of the farm, and lastly those who though highly skilled had no continuing connection with the family and the farm. This order of precedence was given form on such occasions as the harvests which brought a farm's various workers together.

Of the servants one was the chief servant (*gwas mawr*) with the care of the stables as has been said above. He cleaned out the stables unless he had the second servant to do so for him. (Servants were known as the chief servant, second servant, third servant, etc., according to the size of the farm's staff.) If there were so many sons at home that no servant was employed then the eldest son filled the chief servant's position. Very rarely a labourer performed the duties of a chief servant, in which case he was known as *gwrwas*, the same term as was applied to a servant who had entered into a farm by marrying the farmer's daughter (see p. 75). The chief servant fed the horses and did work that demanded a horse and cart. Though someone might be a chief servant while still in his early twenties he took precedence over the other members of the farm's staff. At meal times he was served first, and at those meals during which the food was placed on the table in a pot oven, he helped himself first followed by the other servants according to their ranking, and then by the labourers. When people used their own clasp knives at a meal everyone was expected to finish and rise when the chief servant shut his knife, however brief a time the last to be served had had to eat his meal. At the harvest times the chief servant led the row of scythesmen followed again by the other servants in their order of precedence, and then the labourers. The chief servant decided when to stop to sharpen the scythes and when to resume work. With each November hiring the other servants sought opportunities to improve their positions in this hierarchy of ranks; they and the chief servant alike might seek better farms.

Maids were engaged by the year, and lived in at each farm, sleeping in the house. On small farms the mistress and her daughter did the dairying work, but unless there were daughters of an age to help, one of the maids, the senior maid, helped with the butter-making and in preparing the household's food. There were usually two maids all told and a petty maid besides, but the second maid had no part in the work of the dairy. Butter-making was a task that the maids desired for proficiency in that work improved their chances of marriage. The role of the mistress was so important on a farm that any servant who hoped to get a farm of his own wanted a wife who was familiar with the work of the dairy.

The second maid was employed on the farmyard, as was the senior maid for most and perhaps all of her time. Maids helped in the fields only at harvest times. In the days of flail threshing the second maid was known as the 'barn hand maid' (*morwyn llaw sgubor*), for as the senior maid was connected with the dairy the second maid was connected with the barn. In this connection it should be noted that in South Cardiganshire barns were less storage places than work places, housing the threshing floors. It was the 'barn hand maid's' duty to help the thresher by collecting the straw once it was threshed. This part of the second maid's duties ended when machine threshing replaced the flail, but another of her duties continued. It was her duty to attend at the mill when the farm's corn was taken there to be dried in the kiln. The occasion when corn was taken to the mill to be ground was not one on which people congregated at the mill. One of the mill's servants took the corn from the farm and left it at the mill and it was the miller who delivered the flour when his work was done. But oats and barley, though not wheat, needed drying in the kiln before they were ground, and each farm in turn sent a quantity of these grains (known as *cynnos*) to be dried, a process that took the best part of a day and a night during which the corn had to be spread on a rack in the kiln and turned repeatedly. The farmer provided the culm to heat the kiln; the servant mixed the coal dust and the marl and the second maid shaped the mixture into balls of culm. The servant saw to the corn being dried in the kiln, and the second maid accompanied him to the kiln, tended the fire and saw to their food.

Corn drying followed threshing so that in mid winter there was a fire and company at the mill night after night as one farm after another brought grain there to be dried and ground. Sons and servants collected there to gossip and tease the maid, the whole proceedings being known as a *shimbli*. This continued long after machine threshing replaced the flail because as long as oatmeal was required for the farm-house it was necessary to dry oats at the mill. Oatmeal is no longer used; barley is crushed to feed to the animals, not ground, so that drying is no longer necessary. In the past most of the social occasions of the countryside were closely integrated with farm work.

The maids had no tasks to perform for the servants employed at the farm save one. Servants sent their laundry home or made other arrangements to have their clothes washed (usually by a widow or spinster who undertook laundry work), but after the servants had washed their clogs on a Saturday night the maids greased them with a mixture of lampblack and pig fat which was prepared in advance and kept in its pot in the back kitchen.

When a servant married he became a farm labourer unless he was able to get a holding of his own, which would usually be a horse place (or he might become a farm labourer on marrying only to secure a holding later). At the turn of the century tenant holdings were available, little machinery was necessary, and a small farm would have its own work group. The chief problem was to secure a holding, and to this end a servant had to be industrious and to gain a name as a good servant. He had to be thrifty; if a servant bought more than one pair of boots during the year it soon became common knowledge and gave him a reputation for being careless with his possessions. It served him to attend a place of worship regularly for this was part of the 'character' that helped a man to gain a good name. When a holding was acquired many of those who were neighbouring farmers had been the new tenant's friends and associates in those groups of farmers' sons and servants who shared the same sleeping accommodation as they did their leisure hours, and between whom the links might be considerably closer than has yet been indicated, as will be shown below. These neighbours helped, particularly with the loan of any machines required, as did a previous master to whom the newly established tenant had given good service.

On entering a new holding in this way, a new tenant's chief requirement was stock. A servant who had served seven years continuously at one farm was entitled to a heifer, and a maid to a pair of blankets. (Different farms abandoned these practices at different times, but they were still observed on many farms until the Second World War.) On taking a holding a newly married man used the capital that he and his wife had saved before marriage, he hoped for aid from relatives, and he borrowed if he could, but what principally enabled him to acquire stock was that he could buy in auctions on credit while he would later be able to sell to dealers for cash. Thus on entering a holding in October a newly married man could acquire stock on credit at a time of year when farms were selling off stock they could not carry over the winter. His first major outlay would be the purchase of seeds and possibly fertilizers in the spring. He made sure that he had a litter of pigs ready to sell to meet the expense involved. At the spring sales he sold some of the stock that he had acquired, earning his profit on the animals, which enabled him to pay for those animals he had acquired on credit. Auctioneers gave six months' credit (later reduced to three months); they knew the financial circumstances of those to whom they sold on credit and if someone defaulted they repossessed the animals rather than enforced payment. Weathering this initial period meant hardship for a young couple but many succeeded.

The years during which a young couple held a horse place were their most difficult for they had to collect sufficient capital to establish themselves on a farm if they were ever to join the ranks of 'farmers'. It was necessary to make clothes last whatever their condition, to retire for the night before it was necessary to use a light, to ride the work horse with nothing on the animal's back but a sack in order to save the expense of a saddle, and in some cases to scrape the encrustation of salt from the salted bacon in order to re-use it. But there is no doubt at all that many did establish themselves in this way. Time and again in answer to requests for specific details about servants who became farmers informants have answered 'Well, there was my father . . .'. Further these included servants who were themselves labourers' sons, not farmers' sons. While it is no longer practicable to provide comprehensive data on how many servants and labourers eventually found holdings of their own some indication is given by the fact that in 1900 five of the forty-six farms of thirty acres and over in Troedyraur parish were held by one-time servants, and these farms ranged in size from 58 to 112 acres.

In the material presented above what was of chief concern was the way in which farm work was organized, and the hierarchy of prestige that was associated with different types of work and workers. This was followed by a discussion of how the marriage of a farm servant involved him in changing his work, and how a servant or a labourer could establish himself on a farm, which again pointed to the importance of the friendships that arose between young men who virtually constituted themselves an age group of those who worked with horses. Before seeking to see in the next chapter how farm-house practices accorded with the character of the society at large it is necessary to examine one other aspect of farm organization, namely that each farm co-operated with other farms for some of the year's activities. The most important occasion on which farms co-operated was the hay harvest, and this will be discussed.

Two types of hay were grown, that grown from seed sown each year, and the hay which grew on land reserved as permanent grassland (*gwair gwndwn*). People classified this natural growth according to whether it grew on waste ground or in a field set aside for this purpose. Farmyards were generally on a slope, and the dung heaps stood on the yards. Rainwater percolated through the dung and was allowed to collect at the bottom of the farmyard. Occasionally a channel was opened so that the water could run (or be raked) into the field immediately below the farmyard to fertilize it so that it could grow an annual crop of hay. Farms normally had two hay harvests each summer, the first from the seed hay (which was fed to the horses) and the second from the hay that grew in the 'field under the yard', which was fed to the cattle. Large farms might have a third harvest of the naturally propagated hay that grew on steep slopes or on moist ground that could not be cultivated and used in any other way. The first harvest came in June, depending on the weather, and as work was by hand (in the most literal sense for some operations) there was need for a good deal of labour for some time until the hay harvests were over.

In order that the hay should be of a consistent quality it was necessary that it should all be mown as quickly as possible so that it could dry evenly, and as farms co-operated for the hay harvest it was possible to mow all of any one farm's hay expeditiously. In fact there were two occasions during the hay harvest when farms co-operated, first to mow, and later when the hay was dry to bring their hay carts to each farm in turn in order to carry the hay to the rickyards or wherever else the ricks were built. Other work connected with harvesting the hay was done by each farm individually.

There were customary groupings of farms for this co-operative work. They were generally of adjacent farms and whether they were occupied by relatives or not was accidental. These were not exclusive groups that could be pictured as a number of cells, instead each farm was the focus of its own group without any necessity that the farms that co-operated with it should also co-operate with one another. Thus groups of co-operating farms overlapped with one another over the whole countryside. The same farms co-operated year after year. In some groups the order in which farms mowed was decided annually according to the state of the crops at the co-operating farms. This left room for disagreements and sometimes bitter feelings. In other groups the farms mowed in the same order each year. This might be technically less efficient but it left no ground for personal differences.

When mowing began at any one of a group of co-operating farms the scythesmen assembled at first light and mowed till midday when they dispersed each to his own farm and its work, which at this time of year meant tending its own hay. It was not necessary that the farmers themselves attended at the harvest of any farm in the group of co-operating farms but that each sent a man, whether a son, a servant, or a labourer, with his scythe. As the hay was mown women and strong children of the farm's harvest work group who had been summoned there spread the hay in an even layer over the field, walking backwards the whole time so that they did not tread on the spread hay. In the early 1880s this was still done literally by hand and not with rakes. As one farm of 260 acres had a twenty-acre hayfield it is easy to imagine the labour involved, and this was only the first of several times that the hay might be spread during one harvest. Before the end of the decade rakes were used for this work. Here it may be noted that descriptions of haymaking given by those who worked at it during the early 1880s when the hay was still made into haycocks during the process of haymaking are more or less identical with early nineteenth-century recommendations of how haymaking should proceed.

Given fair weather the farmer again gave notice to the members of his work group that in two days' time they would be required to turn the hay. It was the wives and children who were expected, not the men, and the work that they did was what was normally expected of them and did not contribute to the 'work debt' which was paid only at the corn harvest. The women then raked the hay and the men of the farm's staff used pitchforks to gather it into windrows to uncover the ground that it might dry. After an hour or two the hay was again spread over the field's surface. Some days later when the hay was judged fit for carting it was made into windrows again, so that it could be conveniently loaded and carted to the rickyard—and this was the major occasion during haymaking that called for the presence of all who co-operated and worked at the farm.

Herewith an example. Cefen was a farm of 127 acres occupied by a husband and wife who were childless. They engaged three servants and three maids. There were five cow places that were parcels of Cefen, and the five families that occupied them along with one other family set out potatoes at Cefen. They set five or six rows of potatoes apiece so that they owed thirty to thirty-six days of work debt at the corn harvest, while the wives were required to work at the hay harvest too. Cefen was one of six farms that co-operated with one another for the hay harvest, but not for the corn harvest. When the day came to mow the hay Cefen's servants and the scythesmen sent from the other five farms constituted a string of twelve to fifteen scythesmen who started at first light and mowed to midday. They were accompanied by the women and stronger children of the potato-setting group, who spread out the hay in an even layer. For the next few days Cefen's servants would be away during the mornings mowing at the other farms with which Cefen co-operated, returning after midday. Meanwhile it was the female staff at Cefen and the women of the work group, along with any casual labourer hired by the day, or men sent by other holdings to pay a work debt owing for a bull's services, who turned and spread the hay so that it would dry evenly, and this amounted to forty people or even more, for whom the mistress had to provide food which was taken to the hayfield.

When the final day of the harvest came, Cefen's servants were supplemented during the afternoon by representatives from the other farms, who had been mowing elsewhere during the morning (from first light) and the women and children of Cefen's work group were present making a total of fifty or more people. (Photographs taken at the time confirm this figure.) Their work during the day indicates the character of much of the general social arrangements of the countryside.

The master supervised the work in the fields, the mistress was chiefly concerned with preparing food for all who were present. One or more hay carts were present from each farm, each in the charge of a son or a servant. During the day's work the labourers' implement was the pitchfork and the women's the rake. Women and older men no longer capable of heavy work raked together the hay left behind by the other men as they moved hay into windrows and pitched it into the hay carts. Young women joined the men in pitchings the hay into the carts, glad to be away from the company of their parents' generation and ready for the teasing and bantering that always accompanied haymaking. A man stood on the load of hay in the cart to distribute it evenly and safely. In the rickyard labourers pitched the hay on to the rick, where another built up the rick aided by women who stood on the rick to pass him the hay pitched up from below. Eventually when the ricks had settled a casual labourer would be hired to thatch them. It was an occasion for good fellowship (unless and until the home-brewed that many farms provided made some men drunk and quarrelsome), the younger people making the most of the fact that the final days of the hay harvests of a group of farms were among the rare occasions of the year when the men and the women of a number of farms were brought into each other's company. They were days when more conversational liberties than usual were allowed, and while sitting at ease after tea, which was taken in the hayfield, a man might tussle with a maid in full view of the others present, throw her on the hay, and kiss her.[11] This happened to the accompaniment of the bantering of the other harvesters. The whole episode was regarded with amusement, but it also gave expression to the covert challenge between the men and the women which was present on these occasions.

Older people compared the harvest with those of other years, and reminisced about them. No specific working hours were observed at haymaking and by nightfall many men had worked from first light to dusk, and later. What emerges is that on such occasions farm organization was virtually the organization of the whole countryside. The importance of this is better appreciated if it is remembered that haymaking was not one event. During the late nineteenth century it might last intermittently for months rather than weeks. A cottager's diary records that in 1893 the hay harvest lasted from 19 June to 25 August in his locality, in 1894 from 27 June to 13 August, in 1895 from 17 June to 17 September, in 1896 from 16 June to 3 August, and in 1897 from 5 July to 30 July.[12] The import of changes that have occurred since the 1890s (in this area) can be gathered from the popular comment on haymaking sixty years later: 'It's like funerals, strictly private.'

11 This was known as 'foxing' (*ffocso*).
12 Unpublished manuscript in private possession.

IV

HOUSES AND HOUSEHOLD ORGANIZATION

LOCAL idiom included terms for the major divisions of the society, namely 'great people', 'farmers', and 'the people of the little houses'. There were also terms which constituted a classification of houses—*tŷ dau ben* (a 'two-ended house'), *tŷ singl* (a 'single house', which term does not refer to physical isolation), and *tŷ dwbwl* (a 'double house'). A 'two-ended house' was a two-roomed cottage with or without a half-loft above. A 'single house', or a 'house with a loft', was one in which the two or three principal rooms on the ground floor were arranged in one row, with a similar arrangement on the first floor. Commonly the ground floor rooms were supplemented with a lean-to at the back, which in the case of a farm-house contained the dairy (as may be seen from the examples shown in Fig. 8). In a 'double house' the principal rooms on the ground floor instead of standing end to end were arranged so that there were front rooms and back rooms without a lean-to. Usually there were two rooms in the front part of the house and two or three in the back, as is illustrated by the details in Figure 8. It was rare for a small farm to have a double house but quite common for a large farm to have a single house, indeed many of these were once the homes of the minor gentry of South Cardiganshire to whom reference has been made in an earlier chapter.

The term 'people of the little houses' was concerned with a classification of people, not of houses, and it referred to those who lived in houses without land or with only a field attached. In fact most of the 'people of the little houses' lived in 'two-ended houses', but some had more substantial homes. The élite of the craftsmen were the blacksmiths and the carpenters. No blacksmith lived in a 'two-ended house', indeed blacksmiths usually had substantial stone-built houses. Carpenters too lived in houses which were larger than 'two-ended houses', but they and the blacksmiths alike belonged to the category 'people of the little houses' as will emerge more clearly below. No farm-house was a 'two-ended house', so that those who lived in single houses and double houses included farmers, some craftsmen, and others too, such as schoolmasters, ministers of religion, and the traders who sold seeds, meal, and farm implements as well as groceries.

The second half of the nineteenth century saw an extensive rebuilding of dwelling-houses much as has the period since the end of the Second World War. No attempt will be made to date the earlier period of rebuilding precisely, but it may be indicated that it generally lasted until the 1880s or 1890s, and on some estates into the early years of the twentieth century. Afterwards there was comparatively little major rebuilding (as contrasted with the installation of piped water and electricity) till the Second World War ended. Structurally those few farm-houses which have not been modernized since 1945 are much as such houses were after being built anew or rebuilt during the second half of the nineteenth century. Some may still be seen (1960) each with its open fireplace covered by the chimney's louver which forms a canopy in the kitchen and occupies much of the space in any bedroom above the kitchen.

It is stressed here that this refers to the farming population of South Cardiganshire. In contrast the renovation of old houses and the building of new ones during the nineteenth century continued in the coastal villages after farm-house modernization had ended, which suggests that the agricultural depression of the time affected the pastoral west of Britain more seriously than some authorities have indicated. Here it is proposed first of all to describe the cottages which were the homes of most of the locality's population before turning to examine farm-house practices and to see how these contribute to an understanding of the society with which this study is concerned.

The two rooms of a cottage ('two-ended house') were known as the 'upper end' which was the kitchen and living-room, and the 'lower end' or parlour. The only door opened directly into the kitchen and it was flanked on the one hand by the wall separating the 'upper' and 'lower' ends of the house, and on the other hand by a wooden partition as may be seen in Figures 6 and 7. The wall and the partition between them formed a short passage leading into the living-room. There was one fireplace only which was against the gable in the 'upper end' of the house, and it was during the nineteenth-century period of house improvement and rebuilding which was referred to above that grates were commonly installed. Until that was done turf and wood fires burnt upon the floor, but once grates came into use culm was needed for fuel. However, the installation of grates did little to change the kitchen implements or the cooking methods, for they were installed under the open chimney and its canopy, with a link or crane positioned so that a pot or a

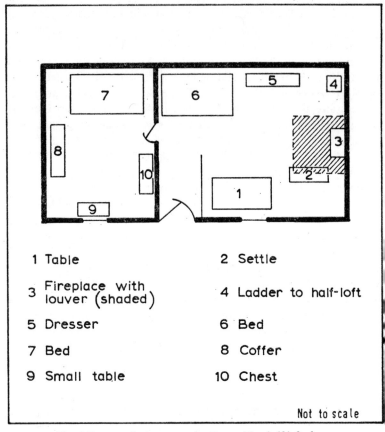

1	Table	2	Settle
3	Fireplace with louver (shaded)	4	Ladder to half-loft
5	Dresser	6	Bed
7	Bed	8	Coffer
9	Small table	10	Chest

Not to scale

Fig. 6. Layout of a two-roomed cottage with a half-loft above.

cauldron could be suspended over the fire, as was the case when fires burnt upon the floor. The oven was always a wall oven which was not heated by the fire in the fireplace, and which might not even be installed in the house itself but in an earth bank outside.

The focus of the room was the hearth while the table stood on one side where the light from the window fell upon it. It was so placed that the partition at the entrance to the room formed a backrest for those who sat at one side of the table. A dresser stood against one

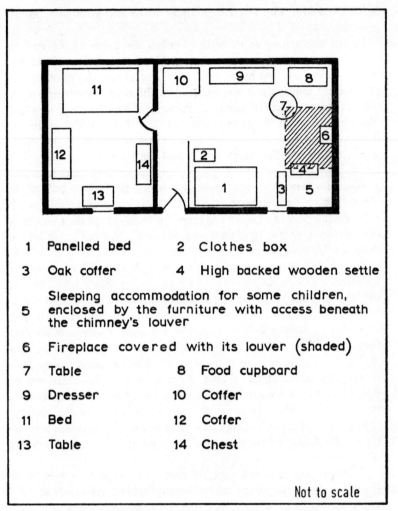

1   Panelled bed   2   Clothes box

3   Oak coffer   4   High backed wooden settle

5   Sleeping accommodation for some children,
    enclosed by the furniture with access beneath
    the chimney's louver

6   Fireplace covered with its louver (shaded)

7   Table   8   Food cupboard

9   Dresser   10   Coffer

11 Bed   12   Coffer

13 Table   14   Chest

Not to scale

Fig. 7. Layout of a two-roomed cottage without a half-loft above.

wall, carrying some of the items of crockery, jugs, plates, and orna-
ments received from different people on special occasions, as for
helping them at a difficult time, or in memory of some deceased
relative or friend. Where there were children in the house the
kitchen had to hold the bed in which the parents slept (usually a
curtained bed but sometimes panelled and fitted with sliding doors),
and its was from the kitchen that the ladder ascended into the

half-loft above. The boys slept there while the girls slept in the parlour.

Unlike the arrangement of the kitchen the focus of the parlour was a table placed centrally or against the window. It bore the family Bible, testaments, and hymn-books for regular use, and photographs of relatives unless these were placed in the corner cupboard along with the more expensive items of crockery. It was the room in which 'the best things of the family' were kept, and into which callers of high standing were shown. The parlour contained too an oak coffer where the bed-clothes were stored, and the room was used as little as was possible while the occupants' activities centred on the kitchen.

When farm-houses were renovated during the second half of the nineteenth century they were rebuilt so as to meet certain needs that pressed on all farms, and to meet the standards of the times in so far as the owners' circumstances allowed. Not all farm-houses needed rebuilding to answer these needs and satisfy these standards, and such farm-houses remained unchanged. (Several of them can be dated with certainty to the later eighteenth century and were not substantially changed until the years that followed the Second World War.) But other farm-houses were altered structurally. No plans survive of their layouts before alteration, but what was said of some of them was, 'You went through the door of the house and there were the cattle's heads *in front of your eyes*'. (The italicized words are a literal translation of the Welsh idiom used by informants and do not imply that the cattle were literally facing whoever entered the house.) What this suggests is that the dwelling-house was entered through the byre, both being under the same roof, and that it was this type of farm-house that was structurally modified during this period of rebuilding.

As milk itself was not sold but made into butter for sale to the dealers a dairy was essential for butter-making. As much of farm work was done by hand rather than by machinery there had to be arrangements to cater for a relatively large staff. These needs were common to all farms. In the outcome there was a measure of standardization about the layout of farm-houses by the end of the years of renovation that have been mentioned. But the people who built these houses anew and lived in them were members of a community that has been in part described and this too was to be seen in certain features of farm-houses and in the way they were occupied and used.

In most farm-houses there was one room which was known as

'the room of the table', or 'board room' (*rwm ford*), which was a mess room. Examples are to be seen in Figure 8. As these examples show, the mess room was usually a relatively small room to which there was no direct access from outside the house. Instead it was entered through the back kitchen and sometimes there was a window

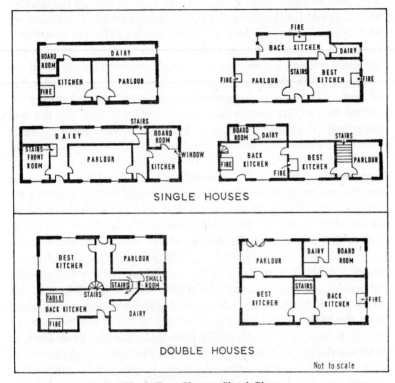

Fig. 8. Farm Houses: Sketch Plans

in the interior wall between it and the back kitchen, a feature of one of the sketch plans presented in Figure 8. Nor was there always a door to this room but simply an opening which could not be closed. It was a bare room and its chief feature was the table with wooden forms or benches standing at its sides. The room contained nothing else unless the flour chest of the cheese press was kept there. It did not have a fireplace.

If there was no 'board room' then the table stood in the back kitchen (in which case this room might be known as the 'board

room'). The table was placed near to the wall with wooden forms
or benches of wood attached to the wall to accommodate those who
sat at the table which was scrubbed clean for use. The centre of the
floor and the space around the fireplace was left clear, for this was
a work room used not only for preparing food but for any indoor
work that was done during the evenings, such as making the wooden
pegs required for thatching the ricks, and for such occasional work
as plucking poultry to be sold at the market. The fire burnt in a
grate installed under the open chimney and the whole fireplace and
the space adjoining it were covered by the mantle that funnelled the
smoke into the chimney. As in the cottage so in the back kitchen
of many a farm-house the grate had only been installed during the
second half of the nineteenth century and until then the fire had
burnt upon the floor. Mounted at the fireplace was an iron crane or
a link, on which pots were suspended above the grate as they had
earlier been suspended above a fire burning on the floor. Not uncommonly a hot-water boiler stood on one side of the grate and a
small oven on the other, but this was not used for cooking but for
drying firewood. Bread was baked in a wall oven which might be
mounted in the wall of the back kitchen or of an outhouse or even in
the earth bank outside the house, whilst most other foods (fresh
meat included) were prepared by boiling in pots suspended above
the fire. Both sides of the fireplace were flanked by wooden settles
or benches, and the servants who slept in the stable loft had the use
of the back kitchen once the evening meal was over. Contrasted
with the back kitchen was the best kitchen, which was not a work
room. It was more comfortably furnished than the back kitchen,
and was the personal room of the master and mistress of the house.
It has already been seen that there was a considerable number of
people connected with each farm, its regular staff, those who constituted its harvest work group, and periodically people from those
farms with which it co-operated at the hay harvests, potato lifting,
and at threshing times. If the usual practices are examined in regards
to the use made of the different rooms of the farm-house some
further understanding of the nature of this society may be gained.

The first work of the day was the care of the livestock, which was
followed by breakfast. The maids milked and fed the dairy cattle,
while the chief servant cleaned the horses, fed them, and he himself
or the second servant cleansed the stables. The labourers and any
other servants tended the store cattle and the foals if there were any.

Breakfast was ready by about seven o'clock and consisted of tea, bread and butter, and cheese. The master and the mistress ate in the best kitchen, the master first and the mistress when everyone else had finished. The servants and the labourers ate in the 'room of the table', or in the back kitchen if there was no 'board room'. So did the children once they were old enough to need no help with their food, and they continued to do so whatever their age until they themselves became masters and mistresses. This was so at all farms. Even at a large farm of over 200 acres which also had a saw mill and other business enterprises attached, and where twenty-four people took their meals regularly in the 'board room' the children ate with the servants and the labourers. The men took their meal first, the women afterwards. At many farms breakfast was followed by household devotions which consisted of a short reading from Scripture followed by a prayer. For this the master left the best kitchen and joined the others in the 'board room' where he himself conducted the devotions, unless the company included an elderly servant or labourer known for his fitness and ability at prayer in which case he was called on to pray. Devotions were conducted every morning but not in the evenings except on Sundays when they were held after the evening meal which was served when people had returned from chapel and church. The women were present but took no part in the devotions apart from reciting verses which were called for only on Sunday nights.

Most farms provided 'ten o'clock food' in mid morning, which the petty servant took to the men at their work places, where it was eaten in the shelter of a hedge. Generally it consisted of tea, bread and butter, and cheese, but at 'good places' some of the liquid was lifted from the broth to be provided at midday, and this was served in mid morning. The petty servant or petty maid again called the men to the midday meal which was taken at noon at farms where no 'ten o'clock food' was provided and some half an hour later where it was provided. The quality and abundance of the food were among the most important features that led to the acknowledgement that some farms were 'good places' while others were not, and servants remained at 'good places' year after year while they came and went regularly at other farms. At mansions bells were sounded to summon the outdoor staff to their meals, while some farms sounded cow horns instead of sending a petty servant to call the men.

Again the master and the mistress ate in the best kitchen and the others in the 'board room' (or the back kitchen). The first course consisted of the liquid from the broth served in wooden bowls and eaten with wooden spoons. The meat (usually salt bacon or salt beef) and the potatoes and other vegetables from the broth were served next on wooden trenchers, followed either by dumplings cooked in the broth, or more commonly by rice pudding which many of the 'poor places' left unsweetened. It was only at harvest times that the main meals were taken out of doors. The maids then carried the food to the fields in wicker baskets, and this was the one time that the oatmeal junkets, which had been a staple long before potatoes were widely eaten, were regularly served. Wherever the meal was served it was the chief servant who decided when it was time to restart work and for each man to return to his task.

An 'afternoon tea' (*te prynhawn*) of tea, bread and butter, and cheese was taken to the fields, after which the men continued with their work. Meanwhile the petty servant fetched the milking cows from the fields for the maids to milk and round about six o'clock the men released the horses from their harness and led them to the stables or to the field in which they were to pass the night if it was summer. The evening meal was taken about seven o'clock, with the master and the mistress again in the best kitchen and the others in the 'board room'. The meal consisted of reheated broth, or the left-overs of the day, and afterwards the servants were free from any further duties.

Two features of farm organization which have previously been noted emerge again in a different way in these details of farm-house practice, firstly the distinction between the master and the mistress on the one hand and the rest of the staff on the other, and secondly the equivalence of sons and servants in the organization of the farm's staff. Sons did the same work as servants, shared the servants' sleeping-quarters, mingled with the servants in their leisure time, and also ate with them in the 'board room'. This arrangement made the subordinate role of the children in farm organization prevail over their status as members of the family. Only when no outside labour was employed did the whole family eat together, and they did so in the 'board room'. On the other hand when a number of siblings had jointly succeeded to the occupation of the farm on their parents' death they all ate in the best kitchen while the rest of the staff had their meals in the 'board room'.

Several of the other distinctions recognized in this society also emerge in practices connected with the farm-house. At home a farmer's son ate in the 'board room' but if he called at another farm at a meal time he was invited to join the master and mistress in the best kitchen as would a farmer visiting the house. A farmer's son had a status in a farming family as well as a subordinate place in farm organization, and it was this that was recognized when he was invited to share a meal away from his home. He belonged to the 'farmers' and not to the 'cottagers'. A farmer's son, one of the six children of the tenant of one of the petty places in the Vale of Troedyraur, recalled that when he called at a neighbouring freehold farm of over 200 acres he was always invited into the best kitchen for he was a 'farm son'.

Recognition of the distinction between farmers and others who worked on farms was given too on those rare occasions during the year when several farmers came together at one farm in the course of their regular work. Farms co-operated at the hay harvest, but each farmer was then concerned with his own hay and he sent deputies to help at those farms with which he co-operated. At threshing time, however, when the farms which co-operated at the hay harvest co-operated again, the farmers themselves went to the farm where the threshing machine was engaged. Threshing took place when the harvest was over and the pressure of work had eased. The time of the year had come when work did not press from day to day and people could relax and take comparative ease. No two farms threshed on the same day for each had to await the arrival of the threshing machine. Threshing was an occasion for companionship which farmers attended personally, and sent their servants too, so that frequently there were more present than were required to do the work. And as the larger farms needed to thresh twice each year there were opportunities for a good deal of visiting among those farmers who co-operated in the work.

Before the first traction engine in South Cardiganshire was purchased about 1905 the threshing equipment was moved from farm to farm by horses. The equipment was in two parts, the boiler and the threshing drum. One part was drawn by horses from the farm which the threshing team was leaving and the other was moved by horses from the farm to which the team was going. The parts were heavy and sometimes needed manœuvring into and out of awkward positions. As the horses which had to do the work were the chief pride of each farm's staff the occasion resolved itself into an open competition of intense interest to all concerned. It was not unknown for a farmer who saw someone else's horses succeed where his had failed to be so offended that he would no longer speak to the owner of the successful horses. Whatever happened other people commented on the horses and on their performance.

Each farmer provided the coal required to heat the boiler and on the morrow of the day when the machine arrived at a farm, work did not begin early for it needed time to carry sixty-odd gallons of water to the boiler and to raise sufficient steam pressure to work the machine. The actual work of threshing involved a number of operations and the way in which these duties were allocated was part and parcel of both the labour and the social arrangements of the countryside. Farmers' sons or servants stood on the corn stack to pass the sheaves to whoever had the task of undoing the binding round each sheaf (or cutting the twine if the crop had been harvested by a self-binding machine). This required nimble fingers and might be the work of a maid, who then passed on the corn to those who fed it into the threshing machine. 'Feeding' was a heavy job. It was work for sons and servants from other farms. The farm's own staff carried the full sacks of grain to their storage places for they knew where to place them, the farmer carrying an occasional sack to see how the storage was proceeding. A servant changed the sacks under the machine and an elderly labourer carried away the chaff which was kept to provide bedding for the calves (and any filling needed for people's beds), a light but dirty job.

Farmers gave a hand with moving the straw which was made into a rick, with an experienced labourer responsible for building it. They gave a hand too with passing the sheaves to those who fed them into the threshing machine, but any farmer would be unwilling if he was asked to do what was considered to be the work of a servant or a labourer. Much of the farmers' time was spent chatting. If rain hampered the work they gathered in the barn to gossip while the others were sent to whatever work was found for them. At meal times the mistress stood at the door of the house inviting the farmers to come to the best kitchen but directing sons, servants, and labourers to the 'board room'. One of the characteristics of the society was that the organization of work was at the same time an ordering of social relationships, and this among people who were in continual contact not only at their work but in their leisure hours in situations which were not directly connected with their employment.

When labourers from other farms were present at meal times they were offered meals in the 'board room'. This too was where craftsmen ate if they were present at meal times. The élite of the craftsmen were the blacksmiths and the carpenters, as has previously been mentioned. Blacksmiths were independent people with a pride in their craft and they were widely known, especially if they made ploughs for use in ploughing competitions. In their smithies they did not burn ordinary domestic coal but a different type of coal which was delivered free from the railway station to the smithy by those farmers who patronized each particular smithy. When, in the early years of this century, T.E., who owned and farmed one of the best and largest holdings in the locality, a farm of over 200 acres, refused to deliver coal to a neighbouring smithy he was promptly turned away, told to take his work elsewhere, and no more work was accepted from him until he had consented to help deliver the 'smith's coal'. Blacksmiths rarely worked at farms for in the nature of the case it was usually necessary to take the work to the smithy, but when they did visit farms to install pipes or pumps, they ate in the 'board room'. So did all other craftsmen and this placed them with the labourers in the category of the 'people of the little houses' and not among the 'farmers'.

It was quite common for boys, on leaving school, to become petty servants on farms for a few years, and then to be apprenticed to learn a craft. The composition of these sibling groups, too, indicates that the craftsmen belonged with the labourers to the 'people of the little houses':

(1) Tailor, dealer, blacksmith, mason, mason, labourer's wife, labourer's wife.
(2) Labourer, labourer, labourer, labourer, mason, tailor's wife, mason's wife.

It was the 'people of the little houses' who set out potatoes in farm fields and by paying a work debt for doing so constituted the farms' work groups. These were the people who ate in the 'board room'. On the other hand those few who did not set out potatoes (or who engaged a labourer to pay the work debt if they did set out potatoes), such as the auctioneer or minister, were invited into the best kitchen.

As the end of this discussion of farm organization and farmhouse practices is now at hand it is as well to point out their salient features in so far as they concern the society that is under consideration here. It can be seen that in any locality farmers constituted a group *vis-à-vis* those who were employed at the farms. It can be seen too that farmers' sons had the status of servants in the organization of farm work, which corresponded to their subordinate place within the family. In the general organization of farm work the sons and the servants were associated with the horses and it was high praise for any farmer to be known as a 'man of horses' (*dyn ceffyle*), which did not so much mean 'horseman' as a man who was known for his personal capacity to assess horses and to care for them. The women on the other hand were associated with the dairy cattle. It appears that according to an old practice which had almost disappeared by the late nineteenth century it was the eldest daughter who named the calves. When they were turned out of the calf box to be kept henceforth with the store cattle she offered each a handful of hay sprinkled with salt, and named the calf. Cattle invariably had Welsh names, while most horses bore English names.

Not only did the organization of farm work assign quite distinct duties to the men and the women with the prestige work belonging to the men, in the case of the men it emphasized the distinction between married status and unmarried status. It provided that cultivation should be the work of the unmarried men, and that labouring should be the work of the married men, rather than that both should do the same work. There was always something anomalous in the position of the elderly servant whose work linked him with the young men rather than with his married contemporaries. Finally while sons had the status of servants in farm organization, they also had their places as members of the family. This was recognized when they visited other farms at meal times and it was recognized too in the practice whereby they were referred to as 'farm sons'.

V

FAMILY AND FARM: (1) YOUTH AND MARRIAGE

THUS far this study has been concerned with the way in which the form of the society was related to the needs of working the land, with the way in which people were arranged into potato-setting groups, and with the social relationships associated with this particular ordering of the society. It has been further concerned with the customary organization of farm labour and with the social roles attendant on this ordering of the work of the countryside. It is now possible to attend to the individual farm. Each farm was the home of a household which might or might not consist of a nuclear family of parents and children. Despite the fact that it was commonplace in South Cardiganshire for a family to occupy the same farm for only a relatively short period, each family was (for reasons to be considered below, p. 118) closely associated with the farm that it occupied. People were and are known by their Christian names or surnames with the names of their farms appended, or simply by the names of the farms. In ordinary conversation people may ask 'Was Coedcoch (a farm name) there?' or may say 'Brynglas was there too', referring to those who farm those holdings. Any inquiry about the history of a particular farm is answered not in terms of cultivation technique or stocking policy but as an account of the family or families that have occupied the farm. When asked about specific farms I always received such replies as 'Such and such a family lived there; they had a son at such and such a farm' or 'They're a family from such and such a place, their daughter married so and so, and there's another daughter married somewhere in the north of the county'. Siblings may be referred to collectively as 'the children of such and such a farm' (the parental holding) throughout their lives when they and their parents have all long left that farm.

But a farm was not only the home of a family. The people who lived on the farm also constituted a unit of production and their resources as a family on the one hand did not necessarily answer the requirements made of them as a unit of production on the other. As a unit of production they were organized in an institutionalized way in order to cultivate the fields and rear the livestock, and they were faced too with the necessity of meeting the requirements of capital and labour which were necessary if the land was to yield the return expected from it. But as a family, those who occupied the farm had a cycle of their own which was independent of the needs of the land, a cycle which included the birth of children, their growth to maturity, their marriages, which in this community involved their departure from the farm, leaving the parents with progressively fewer children at home with them. Thus the size and composition and requirements of the family varied over time, as they did from one farm to another. The requirements of each farm organized as a production unit varied over time too, with the introduction of machinery and with changes in what might profitably be grown and reared on each farm consequent on changes in the economic position of the country at large, and of government policy. But these variations have been independent of changes in the composition and requirements of the family that occupied the farm. It will be necessary to consider how variations in the composition and requirements of the family on the one hand and the requirements of the farm on the other, have been accommodated to one another. Here it may briefly be indicated that this end was accomplished in two main ways, firstly by equating sons and servants in the organization of farm work, in the manner that was discussed in Chapter III, and secondly by a practice of movement from one farm to another as suited the various phases in the existence of any particular family.

It is convenient to open a discussion of this second procedure by presenting a number of case histories which illustrate the movement of families between farms of different sizes and requirements. This, as has been said, was one of the ways in which a family adapted itself to changes in its own constitution, and the examples will help to illustrate the diversity of these circumstances and the variety of ways in which different families tried to accommodate themselves to the different calls upon them. It will then be possible to comment on the significance of this regular movement of families from one farm to another.

1. PENBRYN FARM (46 acres)

Coedcoch, a fifteen-acre holding which is within easy walking distance of Penbryn, was occupied by Sim Evans, his widowed mother, and his brother. Sim Evans worked as a casual labourer and a haulier, for he could keep a horse on a fifteen-acre holding. Thus there were two sources of income, from his work as a labourer and haulier, and from the holding which was worked by his mother. This enabled them to accumulate sufficient capital to move into a neighbouring place of forty-six acres, Penbryn. Soon after the mother died the brother went to work in a shop in the nearest market town and Sim Evans got married. When Sim Evans had moved from Coedcoch to Penbryn, one David Davies moved into Coedcoch. (Fig. 9 is intended as an aid to following the text. People named in the text are referred to in Fig. 9 by their initials. The letters dd refer to the David Davies mentioned above.) He was the eldest son of Rees Davies who farmed Gwndwn (113 acres) which adjoins Brynglas (110 acres), farmed by Rees Davies's brother, David Davies. (Thus there were two David Davieses, uncle and nephew.) Both Gwndwn and Brynglas were parts of the Bronwydd estate, which was owned by Sir Marteine Lloyd. David Davies (the nephew) had moved into Coedcoch on his marriage to the daughter of a neighbouring farmer David Hughes of Dole (52 acres), or to present the same data in another way he was enabled to marry when this holding, suitable for a newly married couple, became available. He and his wife then proceeded to save money as Sim Evans had done before them. While at Penbryn six children were born to Sim Evans and his wife and when the elder of them grew old enough to work, he sought and found a larger farm, Penclos (111 acres). When he vacated Penbryn, David Davies followed him into that one-pair place. Eventually, when all his children were of working age, Sim Evans moved to a still larger farm (193 acres), and then when some of his children married, back to a smaller farm (109 acres). (All these holdings were tenancies, the farms being parts of several estates.) These events occurred during the 1880s and early 1890s, a period of severe agricultural depression. In 1896 two things happened that are relevant here. David Davies of Brynglas (the uncle) went bankrupt and had to sell out; at the same time Dole (farmed by David Hughes, father-in-law of David Davies the nephew) was purchased over the head of David Hughes when the estate was sold.

Thereupon David Davies the nephew, having children of working age, took the tenancy of Brynglas when his father's brother vacated it, and when David Davies (the nephew) vacated Penbryn he was followed there by his wife's father who had to leave Dole because the new owner intended to occupy it. David Hughes was

a widower, and he moved to Penbryn with his youngest son David
Thomas Hughes and two unmarried daughters. On his death (in
1910) his son succeeded to the tenancy; within a few years both
daughters were married as was David Thomas Hughes himself.

Later David Davies of Brynglas decided to retire from farming
having already established his children on farms. Thereupon he was
followed at Brynglas by his wife's brother, who was David Thomas
Hughes of Penbryn. Yet another David Davies then entered Penbryn,
being a cousin of the previous occupier's wife. Eventually the
daughter of David Thomas Hughes (now of Brynglas) married a

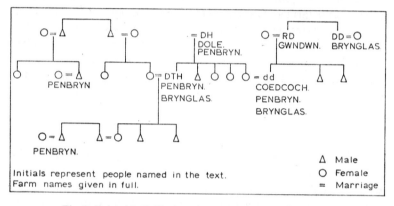

Fig. 9. Related individuals and the farms that they occupied.

man whose brother was married to the daughter of the occupier of
Llechwedd farm, which adjoins Penbryn. During the depression of
the 1920s and the 1930s the occupiers of Llechwedd farm purchased
Penbryn, whose land was henceforth farmed as a part of Llechwedd
and Penbryn disappeared as a separate holding. It is worth noting
certain features that will recur time and again in the cases to be
presented.

1. That as has already been said there are regular movements
   from one farm to another to suit, take advantage of, and cater
   for changing family circumstances.

2. That the people who thus accommodated themselves to
   changing circumstances were 'near kin' to one another in respect of the genealogical distance between them. (The term
   'near kin' is used to refer to ego's spouse and children, ego's
   siblings and their spouses and children, his parents, his

parents' siblings and spouses and their children, his grandparents and grandchildren, as explained in Chapter VII.)
3. That this was only possible granted the readiness of the landlords (who were the local gentry) to grant tenancies to 'near
   relatives' of tenants rather than to others.
4. That the 'family farm' is a rather nebulous notion if it is taken
   to mean a farm handed down regularly from parent to child.
   It remains a useful term to describe a farm worked by a
   family rather than by outside labour.
5. The effects of nation-wide changes in the condition of agriculture namely agricultural depression resulting in bankruptcies and sales of estates, are clearly to be seen.

2. DOLWEN (144 acres)

Dolwen was farmed by one Sam Francis who had moved there
from a seventy-nine-acre farm which he had entered on marrying
the farmer's daughter. Meanwhile Llwynon (220 acres) which
adjoins Dolwen was farmed by a family named Bevan who had
moved there from another part of the county. They had been
evicted from their holding after voting for the Liberal candidate in
the election of 1868, and moved to a petty place on being promised
the tenancy of one of the farms constituting an outlying part of the
Gogerddan estate in the south of the county. This proved to be
Llwynon, so that the Bevans and Sam Francis were neighbours.

But in 1882 outlying parts of the Gogerddan estate were offered
for sale, including Llwynon. It was purchased over the head of the
sitting tenant, a happening that was generally disapproved of, which
was resented, and which gave rise to ill feeling, by Sam Francis of
Dolwen. Sam Francis then vacated Dolwen to enter his newly
acquired property, Llwynon. The Bevans thus had to leave
Llwynon, but they obtained the tenancy of Dolwen, so that they
remained neighbours of Sam Francis, having exchanged farms.
Personal differences between members of the two families led to a
deterioration in the relations between them to the point where one
took legal action against the other.

Eventually Bevan was succeeded by his son John, but none of
the four other sons was established on a farm in the locality, nor
was John Bevan's son, partly because of the difficulty that a family
of strangers had in finding a suitable farm for a newly married couple
without being members of the sort of body of kin who were in a

position to co-operate in obtaining suitable farms (as in the first example cited).

The importance of this consideration is pointed by the circumstances under which the next tenant entered Dolwen. He was Bryn Daniel of Llwyndrain (80 acres), which is situated some eight miles from Dolwen. His daughter married the son of David Thomas Hughes of Penbryn (of the first example cited) on the occasion when Dolwen (which adjoins Penbryn) fell vacant. But Dolwen (144 acres) was a large farm for a newly married couple, who would have to stock it and employ labour to work it. Thus Bryn Daniel removed himself and his family from his home (of 80 acres) and moved to Dolwen (adjoining his in-laws at Penbryn) so that the newly married couple could have Llwyndrain, which being a smaller farm they could stock the more readily. As they both depended on contributions from their respective families to help them to stock their farm, it can be seen the more clearly how inter-farm movements of 'near relatives' helped to accommodate family circumstances and farm requirements to each other. When Bryn Daniel's other children were eventually married Dolwen was then too big for him and he left Dolwen to take a ninety-six-acre farm near the district whence he had come to Dolwen.

Dolwen, being vacant again, was taken by a newly married couple, the Prices, the wife being an affine of the Bevans who had occupied Dolwen before Bryn Daniel. She was the daughter of yet another adjoining farm, one of the largest in the locality and long occupied by the same family who were better able than the Daniels to help establish a young couple on a large farm. The Prices occupied Dolwen for some five years until they purchased Pengwynt (107 acres), in the same locality. Pengwynt and Cwmdethe adjoin one another, and had once been one farm occupied by the owner. He divided it into two to establish both his sons on their own farms. Disaffection between the sons eventually led to a court case. The son against whom the court found arranged to take a farm in another district, offered his own farm for sale, and then committed suicide. On purchasing this farm the Prices left Dolwen (in 1916), which was thus vacant again.

It was taken by one Dewi Harris, who had to leave the farm that he, a tenant, was farming when it was purchased over his head, but after a few years he was able to leave Dolwen (144 acres) for Bryngole (86 acres) when the occupier was declared bankrupt during the

mid 1920s. Another tenant then occupied Dolwen, who in turn left
when the estate was sold and the farm passed into the hands of an
owner-occupier.

The importance of a body of 'near kin' may again be noted and
also that while people so connected co-operated to their mutual
advantage it was not without stress and dissension upon occasion.
It may be noted too how tenants (such as the Bevans in this example)
might move from one district to another aided not so much by
family connections in a particular district as by the connections provided by being tenants of estates which were comprised not of
compact properties but of widely scattered blocks of farms and of
individual farms interspersed among other blocks of farms and individual farms owned by other landlords. The dispersed character of
the holdings provided connections enabling people to move from
one district to another.

3. CWMTERFYN

Cwmterfyn was a freeholding of thirty-three acres. It was
occupied by a widower and his married daughter. Her husband,
David Williams, worked Cwmcul, a 140-acre farm occupied by his
ageing father. Cwmterfyn passed to the occupier's daughter on her
father's death, and to her eldest son Thomas Williams on her death,
while his younger brother David farmed Cwmcul, the 140-acre
holding previously worked by their father.

Thomas Williams had been trained for a profession and lived
in a neighbouring market town. On inheriting Cwmterfyn he let
the ground as grazing land and placed his widowed sister in the
dwelling-house. He then built a new dwelling-house on an adjacent
cow place, installed his sister there, and reconstituted the farm. It
was then let to a newly married couple who remained there until the
death of the wife's mother whereon they removed to join the wife's
father at his farm. By then the brother of the owner of Cwmterfyn
was advancing in years, he had established all his children except his
youngest son, and his own farm was rather remote. Thereupon he
sold his own 140-acre farm and he and his son removed to his
brother's thirty-three-acre farm which was more conveniently
situated and adjacent to the cow place occupied by his widowed
sister. On his death he was followed there by his youngest son,
while his granddaughter moved to the cow place on his sister's
death. None of his five children succeeded him in the 140-acre

holding he had farmed for most of his life. There is here no such notion of 'keeping one's name on the land' as Arensberg describes in *The Irish Countryman.*[1]

It is noteworthy that such a farm as Cwmterfyn was considered a suitable place for a newly married couple to start farming, and for an elderly farmer to retire to when his children were married. It is to be noted too how a farm might be reduced to a dwelling-house or reconstituted, to answer changing circumstances.

4. LLAINMACYN (39 acres)

Llainmacyn was occupied by the owner, one Owens, who had two sons. The elder (Hugh) left on marriage to farm a two-pair place owned by a 'near relative'. The younger (David) stayed at home and eventually inherited the holding. On his death his widow removed to a cow place which was a parcel of the farm, which enabled the son (William) of the late occupier's brother to move into the farm.

He in turn had two sons, Hugh and David. The elder son left on marriage to farm a one-pair place, while the younger stayed to succeed to his father's holding. The elder son (Hugh) then moved to a two-pair place which became vacant on the death of his wife's father, while the younger son (David) who had inherited his father's farm died relatively young without issue.

Llainmacyn was thus vacant and was let to a newly married couple who stayed there until the husband died. At the same time the previous (deceased) occupier's brother's son was married and he in turn entered Llainmacyn, only to leave when his father (Hugh Owens) died leaving a two-pair place vacant. Another newly married couple then entered Llainmacyn, again the husband died relatively young. The farm was then offered for sale and purchased by a newly married couple.

The regular movement of families from one farm to another may be noted again, and it is worth noting too that on occasion free-holders moved as well as tenants, and to note further that certain types of farms were best suited to newly married couples.

5. MELIN WYNT (80 acres) and SYCHBANT (55 acres)

These were adjacent holdings and estate-owned. One James Lloyd was tenant of both; he farmed both as one unit and occupied Melin Wynt dwelling-house while a labourer was installed in Sychbant.

[1] C. M. Arensberg, *The Irish Countryman* (London, 1937).

James Lloyd had one son, Thomas. When Thomas married James Lloyd moved from Melin Wynt to Sychbant (having given notice to the labourer) so that his son could on his marriage take over Melin Wynt. Thomas Lloyd succeeded to the tenancy of both holdings on his father's death.

Thomas had one son Rhys, and on Rhys's marriage he entered Sychbant while his father remained in Melin Wynt. This arrangement continued until the father's health failed when Rhys moved into Melin Wynt and placed a labourer in Sychbant. But in the mid 1920s Rhys was declared a bankrupt, and at the same time the estate was offered for sale. To this end the land was so allocated as to make Melin Wynt a two-pair place (110 acres) and Sychbant a petty place (25 acres). The new owners did not occupy Melin Wynt but let it. The tenant came from a 144-acre holding, which he had entered when the farm he had previously held was purchased over his head; he now availed himself of the chance of occupying a smaller place. Eventually he moved to another farm nearer to his native district and another tenant entered Melin Wynt, while Sychbant was let to the occupier of a separate farm to provide accommodation for a labourer.

6. PENARE (87 acres)

One David Evans, the son of a farm labourer, was engaged as a farm servant (the difference between a farm servant and a labourer was discussed in Chapter III above) at a farm some five miles from Penare. The occupier of the farm at which David Evans was engaged was also the owner of Penare. David Evans courted the maid of a neighbouring farm and married her when his employer promised him the tenancy of Penare when the farm became vacant.

David Evans entered Penare and stayed there a number of years. When he had six children and several of them were of working age he moved to a larger farm in an adjoining parish. Penare was thus vacant and it was taken by a newly married couple, who like David Evans before them, had been able to fix a wedding date on being promised the tenancy of Penare. This couple occupied the farm until the death of the wife, leaving two young children. Thereupon the widower left the farm.

It was then taken by a widow, her two unmarried daughters and a bachelor son, who had to leave their previous holding when the estate was sold. These occupied Penare until they moved to a farm

of 150 acres. The owner then came to Penare, to be succeeded there
eventually by his daughter and her husband.

7. FFRWDMARI (40 acres)

Ffrwdmari was a one-pair place and was occupied during the
1880s by a widow who employed a farm servant to help with the
work. Neither of her two children remained on the farm and on her
death the tenancy passed to one Nathan Daniel. A labourer's son
and himself a farm labourer at one time he had previously occupied
Trefach, a twenty-two-acre holding. Having accumulated capital
there while working as a haulier, and having seven children, he
moved to Ffrwdmari when it became vacant. When all his children
were of working age he moved to a two-pair place.

Thereupon another labourer's son who had occupied a one-horse
place followed Daniel into Ffrwdmari; and he in turn when he
had several children of working age moved on to another two-pair
place.

Yet another one-time labourer then left a one-horse place to take
Ffrwdmari, and when the estate was sold, it was again a one-time
labourer who entered Ffrwdmari.

These points may be noted:

1. In no case in sixty years did child succeed parent on this
   farm.
2. It has been mentioned above (p. 60) that the range in sizes of
   holdings constituted a 'promotion ladder' whereby labourers
   could on occasion become farmers, and this can be seen in
   this example.

8. TANRALLT (166 acres)

During the late nineteenth century Tanrallt was occupied by its
owner William Morris and his wife, along with their three sons and
two daughters none of them being married. When the eldest son,
James, married, the holding was split into two, Tanrallt (56 acres)
and Penrallt (110 acres, including moorland). William Morris, his
wife, and one son died at Tanrallt leaving the third son, Idris, and his
two sisters to occupy Tanrallt. None of them married.

Meanwhile James Morris and his wife occupied Penrallt and six
children were born to them, three sons and three daughters. On
Idris Morris's death at Tanrallt, his nephew William Morris (the
eldest son of James Morris of Penrallt) moved to Tanrallt. The

second son of Penrallt married and moved to Llainmacyn (example 4 above) when that became available, where he died comparatively young. The youngest son at Penrallt remained at his home and succeeded to his parents' holding. His eldest brother (William) who had moved to Tanrallt married there and had two children. Both left the farming community and after the Second World War the farm was sold to another family.

Again one notes the movement of 'near relatives' from one farm to another to cater for family circumstances.

9. PWLLTRO (24 acres)

Pwlltro was owned by one Evan Hughes, who farmed it in one unit with an adjoining eighty-acre farm, Nantyrergyd, of which he was the tenant. His wife was related to the occupiers of a neighbouring twenty-four-acre tenant farm, Blaenos; they had no children.

Blaenos was farmed by Josiah Morgan, who lived there with his wife and five children, three sons and two daughters. On the marriage of their eldest son, Bob, Pwlltro was reconstituted a separate unit, and Evan Hughes who was elderly, removed to it, vacating Nantyrergyd. Bob Morgan was granted the tenancy of Nantyrergyd. One of Bob's brothers emigrated, and the third and youngest brother, John, remained at home with his parents and unmarried sisters. On the death of their parents these three siblings remained at Blaenos.

When Evan Hughes died at Pwlltro the two unmarried sisters moved there from Blaenos, leaving John in occupation of Blaenos. He then married and to him one child was born, a son, Thomas. When this son was ready to marry, his father (by then a widower) left Blaenos to join his sisters at Pwlltro, and Thomas who was thus left in occupation of Blaenos, married. In later years when some of the siblings at Pwlltro had died, Blaenos was sold when the estate was disposed of. Thereupon Thomas and his wife who were childless moved to Pwlltro to join his surviving relatives.

* * *

In the parish of Troedyraur in 1840 there were thirty-eight farms of thirty acres and over (that is, holdings large enough to support husband and wife without a regular extra source of income) excluding those in the occupation of the gentry and those which disappeared through amalgamations in the following years. During the period 1840–1900 there was continuous succession of parents by

children in the occupation of eleven of the thirty-eight farms.[2] Four
of these were freeholds, four were tenant farms throughout the
period, and three were tenant farms purchased by their occupiers
during the period (though the mortgagees foreclosed soon after 1900
in one of these three cases). The succession of parents by their
children was broken once or more often in the case of the other
twenty-seven farms as families moved from farm to farm and as
children married and moved to occupy farms which had become
vacant when other people moved.

This movement from one farm to another was not a peculiarity
of this parish. A South Cardiganshire squire (H. M. Vaughan) stated
that only one of those families occupying the eight farms comprising
the Llangoedmor estate in 1885 was still a tenant in 1915, and that
family was on a different farm.[3] It is worth noting that as families
moved (of their own choice) to answer their changing circumstances
over a period of time, the occupiers of some farms might not move
at all, such as childless couples whose circumstances did not change
(in respect of what concerns us here). At the same time the common
practice of movement from farm to farm was old established, and,
as numerous individual cases show, not the consequence of the sale
of estates. One example may be cited. John Jones of Blaenannerch
(1807–75), a celebrated preacher, was born at Blaenpistyll Mill,
but then moved with his parents first to the farm of Cytir Bach, then
to Cenllefaes farm on his marriage, back to Cytir Bach when his
father was no longer able to work the farm himself, and to his own
house, Brynhyfryd, finally. It was the case that a minority of families
remained on the same farm for one generation after another, but
those which did occupied a prominent place in people's image of
the community that they lived in.

It is intended to revert later to the significance of the co-operation
between closely related individuals to enable families to obtain
farms that suited their changing circumstances. Here attention may
be drawn to other consequences of this regular movement of families
from farm to farm.

Prior to the introduction of those machines which first reached

---

[2] Sources: Tithe Apportionment Map and Schedule, parish of Troedyraur, 1840;
in N.L.W.: Census of 1861, recorders' notebooks, in the Public Record Office:
Sales Catalogues, in N.L.W.; Electoral Lists, in N.L.W. These last during the late
nineteenth century distinguish between 'occupation electors' and 'ownership
electors', the latter being qualified to vote by the ownership of property within the
parish. The property owned is specified in the lists.

[3] H. M. Vaughan, p. 181.

the area in the last decade of the nineteenth century each farm required a harvest work group. Each farm was also a member of a group of farms that co-operated at the hay harvest and on other occasions. As families moved from farm to farm membership of these functional groups was based on the farm rather than on the

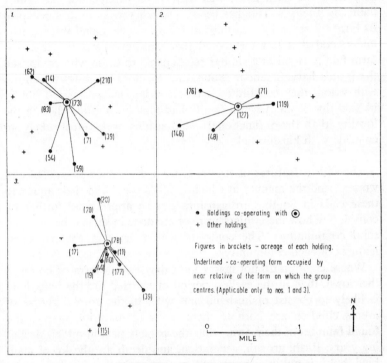

Fig. 10. Farm Co-operation Groups

family that occupied it or on that family's kin links. In numerous cases co-operating farms were in fact occupied by relatives, but the basis of their co-operation was not the fact of kinship. Thus families from other districts came to occupy farms in the area and worked their farms without any difficulties arising because they had no relatives in the locality. The immigrant farmer took over his predecessor's roles in the harvest work group and the inter-farm co-operating group because these were based on the farm rather than on kin. In fact kinship could hardly have provided a general basis for such work groups because the movement of families to suitable

C 7899   E

farms dispersed relatives over too wide an area to allow work groups to be based on kin.

At the same time, as has been noted, people closely associated a family with its farm. In everyday speech family names were connected with the names of the farms that those families occupied; frequently the farm name was used as a synonym for the family name, as it still is. This has been so whether a family has occupied its farm for several generations or for a few years, and this is easily understood if it is remembered that whichever was the case, the farm family stood in a similar relationship to those who comprised the farm's harvest labour group, and to those other farm families with whom they constituted a co-operating group. It need hardly be said that it was when a family had long been established in the locality that these functional relationships were most likely to coincide with kinship links.

*   *   *

The salient features of the way in which farms were staffed and worked, and the manner in which families organized their affairs to these ends in South Cardiganshire can be appreciated further by considering how these things were or are done in some other western rural communities. The contrasts will underscore the particular features of each community.

Where children do not marry until they have a farm or house of their own, then in the earliest period of married life the household is likely to consist of husband and wife. In the normal course of events children are born to them, grow to maturity, marry, and found families of their own, while the parents pass into their declining years. Birth, growth, generation, and death are the lot of mankind, but it is not in one way and one way only that families can be patterned or structured, or can organize their affairs, to meet the universal experiences of life and death. Families are members of communities, and their internal arrangements both contribute to the nature of the communities of which they form a part, and are affected by the nature of the communities in which they are embedded.

Arensberg has described one such set of family arrangements as he found it in rural communities in western Ireland during the early 1930s. He was describing 'small farmers' with farms of thirty to fifty acres, who cultivated 'gardens' of oats, rye, potatoes, cabbage, turnips, and reared calves which were sold off annually. 'Nearly all he (the farmer) raises he consumes at home; his family and his farm animals take the greater part of his produce. It is only his surplus and his annual crop of calves which break out of the circle of sub-sistence and in so doing bring him the only monetary income he receives.'[4] These farmers relied upon family labour, on sons, daughters, and other relatives, and on the help of neighbours when extra hands were needed.

Their dwelling-houses consisted of two or three rooms on the ground floor with or without a loft above. The rooms were disposed in one row, the whole being oriented roughly east-west. One room was always the kitchen, another always the 'west room', and if there was a third room it was a bedroom. 'The west room was a sort of parlour into which none but distinguished guests were admitted. In it were kept pictures of the dead and emigrated members of the family, all the "fine" pieces of furniture.'[5] Younger members of the household could not enter it without their parents' permission.

The farmer and his wife worked their land on their own, with the help of a relative and occasional aid from neighbours until their children grew old enough to work. Then the whole family worked as a unit, the children, even if adults, receiving only occasional pocket money from their father. The critical point in the family's affairs came with the marriage of the son who was to inherit the farm. 'A match usually begins when a farmer casts round for a suitable wife for one of his sons. The son to be married is to inherit the farm. The farmer has full power to choose among his sons. . . . Today the farmer looks forward, ordinarily, to "settling" only one son "on the land".'[6]

Discussions about the 'fortune' that the young woman was to bring with her were carried on through an intermediary, and the arrangements made were that while the young man's father made over the farm and all its appurtenances to his son, the young woman's 'fortune' went not to the young man but to his father. It then provided in part for compensating those children who were not to be 'settled on the land', but it provided chiefly for the 'fortune' needed to wed one of the daughters to another farmer's son. 'Usually only the heir and one daughter are married and dowered, the one with the farm, the other with fortune. All the rest . . . "must travel". Thus either at the match or in preparation for it comes the

[4] C. M. Arensberg, p. 39.
[5] Ibid., p. 24.
[6] Ibid., pp. 72–3.

inevitable dispersal of the farm family. The unit must break up'.[7] Before the nineteenth-century famine clearances and land reforms all the sons and daughters could hope to be provided for on the land, but by the 1930s it was no longer so.

Among the family arrangements made at the time of the heir's marriage was the provision that the west room was to be reserved for the 'old people'. Thus on the heir's wedding, the parents moved into the west room, the young couple into the kitchen, one daughter was 'settled on the land', and those other siblings who were still at home 'travelled', that is, left their home for whatever opportunities there were for them, which in practice frequently meant emigration. Here can be seen one means in which families' ways of arranging their affairs to meet the circumstances of the birth, growth and marriages of the children, and the transmission of the homesteads from one generation to the next, were institutionalized.

Before discussing a second rural community, it may be noted that the above is not only an example of the way in which family arrangements were institutionalized, but exemplifies too a settlement of a family's property that affected all members of the family, a settlement that was made on the heir's marriage. It is worth noting the similarities and contrasts between this arrangement and the family settlements whereby the landowning classes in England and Wales tied up their estates, from the seventeenth century onwards.[8] Here too there was a resettlement of the family estate once in each generation on the heir's coming of age, and usually in connection with his marriage. The Common Law allowed estates to be limited to an unborn son, but not to the unborn son of an unborn son, that is to the first generation yet unborn, but not to the second or still later generation yet unborn, and an estate tail was essentially barrable. Thus estates could not be rendered indefinitely inalienable by direct legal means. Families attempted to gain the same end by resettling estates once in each generation. The heir was customarily the eldest son. After his father's death he would have an estate which he could freely dispose of, but until then he had nothing of the family's property. Hence he would usually be willing to give up his prospective estate in return for some immediate financial provision, followed by a life estate for himself and an entailed interest for his

---

[7] C. M. Arensberg, p. 79.

[8] The last quarter of the nineteenth century saw a protracted discussion of the land-holding system in Britain. See, for example, G. C. Brodrick, *English Land and English Landlords* (London, 1881).

as yet unborn heirs. At the resettlement provision would also be made for clearing debts on the estate, and portions could be allocated to the heir's siblings.

Thus estates were entailed anew once in each generation, so that the family could keep its 'name on the land' as the Irish countryman would have it, and anticipation of the heir's marriage would occasion a settlement of family property affecting all the family's members. There are obvious differences here from the arrangement that Arensberg describes among Irish 'small farmers'; that the heir was customarily the eldest son in the one case but not in the other, that the heir came into the estate on his father's death in one case but on his own marriage in the other, but the critical point in the family's affairs is again connected with the marriage of the heir.

Our second example of how family arrangements were institutionalized in a rural community is that which A. D. Rees delineated as it was in a part of mid Wales during the 1930s and 1940s.[9] He described a farming community in that part of the hill country of eastern Montgomeryshire which lies between the Berwyn uplands on the one hand and the Severn valley on the other, an area better suited to pastoralism than to arable cultivation. Most of the farms ranged from 20 acres to 200 acres in size and for their income farmers depended primarily on rearing cattle and sheep for sale as stores, rather than on fattening for slaughter.

Rees described how newly married couples established on farms would need to employ outside labour to work their farms during the first phase of their married life when they were childless or with young and growing children. But it was the second phase that constituted the ideal that was only temporarily possible. Then with children of working age there was no need to employ outside labour, and families were able to accumulate capital while the children received only their keep and occasional pocket money. They did this knowing that each would receive his or her portion upon marriage. Unlike the Irish farmers of County Clare among whom in Arensberg's account the marriage of the son who was to inherit the farm occasioned 'the dispersal of the farm family', in mid Wales no one succeeded to the farm until the others were provided for. The children were established and left home one by one leaving one child at home unmarried (with a preference for the youngest son), who eventually succeeded to the tenure of the holding. In Ireland the

---

[9] A. D. Rees, *Life in a Welsh Countryside* (Cardiff, 1950).

farm was transferred to the heir upon his marriage and his wife's 'fortune' went to establish one of the husband's sisters on the land. In mid Wales the marriage of one son did not bring a bride's 'fortune' to the young man's father, and there was no transference of the parental home until the other children were established, provided the parents were alive.

In the Irish example the family's property was resettled once and completely, in the Welsh example it was resettled portion by portion on a number of occasions as each of the children married. It may be added that the arrangement that prevailed in mid Wales required that farms be available to those who on marriage left the family holding (if they were to remain within the farming community) so that direct succession of parent by child could apply in only a limited number of instances. In Rees's account of mid Wales the family's circumstances and the farm's requirements were accommodated to one another by the employment of servants in lieu of sons when the children were too young to work and again when they had left the family home. In South Cardiganshire, as has been mentioned above, the divergent requirements of the family and the farm were accommodated to one another in two ways, employed singly or in combination, by hiring servants in lieu of children, and by periodic movement from farm to farm as it suited the family's circumstances during different phases of its cycle. This periodic movement from farm to farm has already been illustrated and the earlier phases of the family's cycle can now be discussed with an eye to seeing how they were connected with and how they affected other features of the community of which each family formed a part.

In South Cardiganshire a newly married couple taking their first farm were (and are) said to 'start their world' (*dechrau'u byd*) at that farm. This did not refer solely to marriage but to the setting up of a new and distinct household in the community. It referred to the establishment of a newly married couple on a farm where for the first time they managed their own affairs. Early in their married life in the first phase of the family's cycle, before they had children of working age, they might as has been noted need to employ servants in order to work the farm. This state of affairs lasted until their own children were old enough to work full time on the holding, which the children could not do until they were of school leaving age. Farm children started to help with the work well before this, fetching the cattle from the fields at milking-time, the boys leading

the horses as soon as they were allowed to do so, while by the age
of twelve the girls' hands were strong enough to enable them to
help with the milking.

By the time boys were fourteen they were entrusted with a horse
and cart, and at fourteen or fifteen years of age they were ready
to undertake their first full day's work as scythesmen during the hay
and corn harvests. Sixty to seventy years later many of them have
recalled that occasion which they remember very clearly on account
both of the muscular fatigue they then experienced, and because
their work placed them henceforth among the men and not among
the children. At fifteen a young lad started to plough, and within
a year he was a member of what was virtually an age grade composed of those young unmarried men who worked with horses.

A farm servant (*gwas*) was quite distinct from a farm labourer
(*gweithwr*). It will be recalled that a farm servant (*gwas*) was
an unmarried and usually young man living in at the farm and
employed to work with horses. What is relevant here is that when
farm servants were employed they slept on the first floor of one
of the outhouses. This was usually the stable loft but it might be
above the byre. In either case the occupants came and went
independently of the farm-house. Within the stable loft each servant
had a frame bed with straw laid on it, and a mattress filled with oat
chaff rested on the straw. (Oat chaff was considered the best bedding
material and was widely used.) Each servant had a metal trunk in
which to keep his belongings. The servants generally shared the
stable loft with the chaffing machine, which was used to chaff straw
when preparing feed for the horses. There was usually a partition
between the sleeping-space and the chaffing-space, but this was not
invariably the case. The servant's day has been described and here
it need only be recalled that the servants used the back kitchen as
part of their accommodation as well as the stable loft.

When a farmer's son reached the age of fifteen or sixteen he moved
his sleeping quarters from the farm-house to join the servants in the
stable loft. He did this whether or not there was room in the
dwelling-house, and far from being a hardship it was the occasion
when for the first time he could come and go free from parental
supervision. (Any son who chose not to move into the stable loft was
considered effeminate by his contemporaries who would taunt him
with it, and he would be the subject of good deal of ridicule.)
Farmers still break into a smile when they recall the time when they

themselves had first moved into the stable loft. While the farm had some sons of working age but still retained some servants, both sons and servants slept in the stable loft. They were very closely associated in work and leisure and they constituted that characteristic group of young unmarried men who roamed the neighbourhoods in small groups after working-hours, or congregated in craftsmen's workshops or at habitual outdoor meeting-places in fair weather. They were the practical jokers of a neighbourhood, removing and hiding gates on November Eve and New Year's Eve, and playing pranks that sometimes called for as much manual work as did their daily occupation. On one occasion the police were called in to investigate the apparent theft of corn at harvest time when in fact a group of young men had dismantled five small stacks (*sopynnau llaw*) of corn and rebuilt them as four stacks. On occasion too these tricks were played on people who had incurred general displeasure, as when the only gate removed in a district belonged to a man who had refused to contribute to a widely supported charity. He understood well enough why his was the only gate to be tampered with.

As the children grew older the family entered the second phase of its cycle during which the children could undertake the farm work themselves, and servants were dispensed with if that was at all possible. For some years there would be few or no wages to pay, for the children were not paid regularly but received small sums of money on particular occasions such as visits to the more important fairs, or to farm auctions in the neighbourhood. This was the time when the family acquired the capital that would be needed to establish the children on their own farms when they married, for it was on and through marriage that each child received his share of the family's resources to which he had contributed some years of unpaid labour. While the family worked as a unit on its holding the father retained control of the farm's management, and of the marketing of stock, which prior to the establishment of auction marts in this area during the First World War involved individual bargaining with a dealer about each animal sold, either on the farmyard or at fairs in one of the market towns in the area. So firmly did many farmers keep the dealing in their own hands that many a farmer's son found himself quite inexperienced and at a disadvantage *vis-à-vis* the dealers when he eventually came to manage his own affairs.

On many farms there were more children than could be employed on the family's farm and these, if they were to remain within the farming community, became servants themselves. One farmer put it simply, 'We were too many children at home so I had to go and serve'. Such farmers' sons were not infrequently employed as servants on farms held by their relatives: for later discussion it may be noted that this had a bearing on the constitution and organization of farm households. It should also be borne in mind when considering the constitution of those groups of young unmarried men who figured prominently in the life of the countryside.

It has been mention that it was when a farmer's son moved to sleep in the stable loft that he was free to come and go without parental supervision. This meant in particular freedom to go courting. Girls were rarely away from their homes (or from the farms where they were employed) in their leisure time except when attending formal functions. After these functions (such as a religious service or a concert) both the youths and the young women formed themselves into several small separate groups of the one sex or the other, in which siblings of the same sex joined different groups as far as that was possible. There followed an institutionalized play by which young men and women paired off once they were out of sight of the adults. A youth would say quietly to a friend in his group, 'Go and fetch so and so for me' (though the groups of young women were within easy hearing distance), and the friend would say to the girl, 'So and so asks if you'll come'. After a good deal of cross chatter while the groups of young men and the groups of young women were yet within easy hearing distance but still distinct groups, pairs would detach themselves, eventually leaving those left unpaired to go their ways independently. Among a group of youths there were frequently partners who acted thus for one another on various occasions, but this was by no means invariably the case.

These occasions apart, courtship took place when all had retired for the night except for the groups of young men who were accommodated in the farms' stable lofts. A youth would initially make an appointment with a young woman, very frequently through an intermediary who would be a friend of, and sent by, the youth. He arranged that the young woman would admit into the house the youth for whom he was acting. This was a frequent arrangement if only because of the difficulty that faced a young man working on one farm in contacting a young woman working on and largely confined to another farm from early morning to late evening. On reaching the house after dark the youth attracted the girl's attention by throwing a handful of small stones or gravel at her window, whereupon the door was opened quietly without waking other members of the household and the young man was admitted either to the girl's bedroom and her bed, or to the farm kitchen. 'Courting in bed' (*caru'n y gwely*) had not entirely disappeared before the end of the nineteenth century, but it is not a matter that people will readily discuss beyond saying that it occurred, and that it was the usual practice to keep a bolster between the couple, which was not removed without the girl's consent. It was common for a 'farm son' (as the Welsh expression *mab ffarm* has it) and a farm servant who shared accommodation in the same stable loft to go courting together to another farm, the one to court the daughter of the house and the other to court the maid. (In this connection it is noteworthy that maids and daughters never shared the same sleeping-accommodation.) Though the composition of the groups of young men changed as the members married, yet close personal ties survived between individual farmers' sons and individual farm servants. These were the ties which were of particular importance should a servant require the help of his neighbours if and when he came to occupy a holding of his own on his marriage, in the manner described in Chapter III.

Much more common than 'courting in bed' was 'a night of watching', *noswaith o wylad*. *Gwylad* (watching) bears many meanings— it means sitting up throughout the night with someone who is ill; in the context under discussion here it meant a night sitting up courting in the farmhouse kitchen when the young man had been admitted after knocking at the girl's window, a procedure known as 'knocking and opening' (*cnoco ac agor*). It would be said of a listless worker, 'He's good for nothing today, he had a "night of watching" last night'. When discussing this it was mentioned that the parents, and more especially the mother, had a pretty shrewd idea as to who was visiting the house by night and that it would be made plain enough to the daughter if her visitor did not receive her parents' approval. In the folk image of the society it is certainly to the mother that oversight of the daughter's courtship is attributed. Courtship proceeded in private, the couple paying no particular attention to one another in public. Their first appearance in public might well see them subjected to a procedure known as *mofyn a tynnu*, 'fetching and drawing', to which any couple might be subjected until they had appeared together on several occasions, and which is described below. In a sense while courtship concerned a young man and a young woman, courtship escapades were competitions between the young men, and nowhere was this clearer than at fairs, and at those wedding festivities, the biddings (*neithiorau*) which, though in decline almost to the point of disappearance at the end of the nineteenth century, merit mention.

The nineteenth-century bidding is to be described in greater detail below. Here it need only be noted that at a bidding the relatives and close friends of the bride and groom and those who were repaying marriage debts (*pwython*) congregated in the farm-house, while other young men and women repaired to the barn where forms had been placed to provide seats, and cakes and drinks were brought in for sale to those present. A youth would pair off with a young woman in the barn, buying cakes and drinks for her, while other youths attempted to draw (*tynnu*) the woman away from him, buying her further cakes in the process of doing so. There resulted a competition in which one man tried to retain the girl while others tried to draw her away. Sometimes the attempt to retain her and draw her away took physical form, one man trying to pull her away by the arm while the other tried to hold her to the point that the woman's clothes were torn and fighting resulted. In any case strong feeling ensued between someone who had 'lost his sweetheart' and whoever managed to entice her away.

A similar practice continued at fairs well after the traditional bidding had disappeared. The first few times a young man and a young woman appeared together at fairs, or if they paired off for the first time at a fair, one of a group of young men was sent to 'fetch' (*mofyn*) the young woman away on behalf of another member of the group. Her suitor had to buy her fairings in order to keep her, while the young man who tried to draw her away had to buy her fairings in order to entice her. (A 'fairing' was a gift bought at a fair, such as a brooch or an ornament.) In the course of an evening a fickle woman might be drawn away from several suitors in turn, in which case she would be loaded with fairings, and one of the first questions asked a young woman after any fair was, 'How many fairings did you have?' Similarly a much-sought-after woman would be loaded with fairings even if she did remain with her suitor the whole evening.

A contemporary account describes the proceedings thus:

> Let us say that Mary Cwmgeist (the name of the farm) is going to Llandyssul fair, and as she is an attractive woman Sam Gwarcod feels that he would like to court her. He would try it thus: he would send Will Moelhedog to her, to fetch her to the tavern to receive fairings. If she went with him, and if there were many who wanted to fetch (*mofyn*) her away, soon Sam Penrallt would come to her to fetch her for Evan Glan-clettwr, and there would be a hard contest for her. Occasionally the youths would fight one another rather fiercely. Now the best man in her opinion was the one that she allowed to accompany her home.[10]

It was to this enticing that a woman was liable to be subjected on first appearing in public with her suitor, but it ceased after she had remained faithful to him on a few occasions.

The families whose activities are here generalized in terms of a 'family cycle' were members of a community in which the role of 'farmer' was held in high regard, and in farming families the parents generally wished to establish their own children on farms if that was possible and they had the resources to do so. To do this was much more important in the late nineteenth century and the early years of the twentieth century than it is today, for since then new roles and occupations are open to individuals and the education that many of them require is now much more widely available.

To establish a young man on a farm involved two things: his marriage, and finding a suitable farm for a newly married couple, the suitability or otherwise of a particular farm being dependent on the resources available to the young couple. While on their parents' farm, as a rule children were expected to remain unmarried, and when both parents died leaving adult siblings at home these too remained unmarried (as in example 8, p. 114) until and unless provision was made for some of them to leave (as in example 9, p. 115). People are quite explicit that marriage and securing a farm were connected with one another. They state that a farmer's son did not think of getting married until he could get a farm of his own, and in the normal course of events the practice bore out the state-ment. As marriage and finding a farm of one's own were inter-connected, the marriages of the children and their establishment on farms constituted critical points in the family cycle as well as major

10 W. J. Davies, *Hanes Plwyf Llandyssul* (Llandysul, 1896), pp. 239–40.

changes in the status of the individuals concerned. Marriage was a critical occasion because the farm a newly married couple were to occupy had to be equipped and stocked from their families' resources, and this affected every member of the families of both bride and groom.

Indeed in a society where a farm family worked as one financial unit in which the adult children received no wages for their labour but expected their dues on their marriages, these marriages could not be such as involved no one but the betrothed couple. They immediately and necessarily concerned the parents and siblings of the bride and the groom, personally and financially. The manner in which any one child was established on marriage affected the provision that remained for establishing the other children, and was thus of direct importance to each and all the children as well as to their parents. Further, as the families of both parties to the marriage would contribute towards establishing them on their marriage, the standing and resources of every member of one family was of immediate consequence to the members of the other family when marriage connected them. Not infrequently (as is discussed below, p. 136) the immediate circumstances of marriage depended on the co-operation of relatives and as people recognized that there was a connection between affines which called for recognition if only in warmth of greeting and general friendly relations, any marriage was a matter of some concern to many other people than the nearest relatives of the bride and groom. The circumstances of any marriage were also of general and remarkable interest to other people who lived in this 'parochial' community. During 1960 I was told several times of a 'great match' that the daughter (one of five children) of a 34-acre tenant farmer made with the freeholder of a 150-acre holding. This marriage took place about 1878, some years before the birth of any of those with whom I discussed the matter. Some of my informants were contemporaries of the children of that marriage and familiar with their circumstances, thus after a lapse of some eighty years the circumstances of a celebrated 'great match' were still widely known.

When a young couple intended marriage the allocation of their families' resources that was necessary to establish them on a holding of their own was in the hands of their parents so that much depended on whether the parents on either side considered the match suitable. In any disagreement between the parents and their children the children were at a disadvantage because when they came to marry they had already worked for several years for no more than occasional pocket money, while they could not gain their share of the family's accumulated resources accruing to them except with the consent of the parents. On occasion courtships were terminated or marriages delayed for years while parents and children remained in disagreement.

Meanwhile though children saw their years of unpaid work as entitling them to provision when they married, many parents saw them primarily as payment due to them for rearing the children (*talu am eu magu*). As will be noted later, the most bitter quarrels in this community were not between kin groups, but within kin groups, between siblings, and one of the situations in which these quarrels developed was an intended marriage when some of the children felt that their own future prospects were adversely affected by the provision made on the marriage of a sibling.

The marriage of a farmer's child therefore was of major concern to the individual members of two families if only because one received one's portion on marriage, not on the parents' death. It concerned others as well, for in so far as individuals recognized and acknowledged their ties with their kin and affines, the status of all of them was affected by the marriages contracted by every one of them in the eyes of a community where that interest in one another's affairs that is part and parcel of 'parochialism' made people sensitive to the opinions of others.

For both these reasons there were difficulties in the way of a marriage between a farmer's son and a cottager's daughter. When this did occur it was said that the man had contracted a 'petty match' (match *bach*). Marriages between farmers' daughters and labourers were not as rare and when such a marriage took place it was usually understood that the best the young woman could expect was that her family would help to find them an auxiliary holding such as many labourers managed to find in any case.

There was one other consideration that affected the demands made upon two families when they were jointly concerned in establishing a newly married couple on a farm, a consideration that was applicable particularly during the late nineteenth and early twentieth centuries. For the greater part of the nineteenth century, as in earlier times, there was in south-west Wales an institution whereby newly married people received financial aid, and this was the bidding (*neithior*), a wedding festivity on which a complex of activities centred. It concerns this study not because it was of major importance at the end of the nineteenth century but because its decline made the obligations of individual families heavier in that period, which was in any case a time of agricultural depression. Without knowing how important the bidding had been there is no way of appreciating the importance of this extra demand upon families.

It is T. M. Owen who has drawn attention to the bidding and shown the sociological importance of what had previously rarely been thought of as anything but an interesting custom.[11] Marriage celebrations included a festivity to which people were bidden by word of mouth or by letter. Hence these occasions were known in English as 'biddings'. The proceedings at a bidding took place partly in the farm-house and partly in the barn or another convenient building. The younger people congregated in the barn, where young men were in open contest for the company of young women (as was noted on p. 127). In the farm-house there assembled those people who had come to render 'marriage dues' (*pwython*), which were contributions in cash or kind that were recorded against the name of each contributor by a clerk appointed for the occasion. In various 'counts' (the records made by clerks) that I have seen, separate accounts were kept of the tea, sugar, bread, butter, and cash received. In addition people bore gifts (*rhoddion*) which were distinguished from the marriage dues, and were presented at the bidding. Some hundreds of contributors were present at many biddings, and as bride and groom might hold separate biddings the result was that there might be a very large number of contributions. There were 195 contributors at the bidding held by a carpenter's daughter at Cwmul near Llandysul in South Cardiganshire in 1825; her husband held a separate bidding at his home. In another bidding held in 1833 near Llanwenog, a few miles distant from Cwmul, there were 369 contributors. In both these cases the names of the contributors and the nature of their contributions were recorded. More general accounts speak of biddings at which there were 500 to 600 people present. Money contributions generally ranged from sixpence to half a crown, sums of one shilling being commonest. At that bidding held in 1883 when there were 369 contributors present (as mentioned

---

[11] T. M. Owen, 'Some Aspects of the Bidding in Cardiganshire', *Ceredigion*, vol. iv, No. 1 (1960), pp. 36–46.

above) the cash contributions amounted to £30. 14s. 0d., and this did not include what was received at the bride's bidding. Such a sum was then sufficient to stock a small holding, so that the bidding was of major importance in enabling a family to establish the children upon farms.

The marriage dues (*pwython*) referred to above were debts for which the debtor might be held liable at law; they were also transferable and heritable. People might 'call in' debts due to them on the marriage of a relative; thus at a bidding held in Llandysul in 1843, those debts due to the groom were called for, and also those debts still due to his parents. In addition further gifts to the groom were invited on behalf of the groom's grandfather, two brothers, and a sister-in-law who undertook to make repayment whenever requested to do so. The bride's parents called for repayment of debts due to them, to the bride, to the bride's deceased sister, and for further gifts on behalf of the bride's brother and sister and three pairs of uncles and aunts. At a bidding held in 1878 it was requested that repayment of debts due to the groom, his parents, his two brothers, and his aunt be made to the groom. It was requested that repayment of debts owing to the following be made to the bride: to the bride herself, her father, her two brothers, and her sister.

The bidding proper was followed by another and comparable occasion, to which people were again invited: this was known variously as 'repairing to the marriage house' (*cyweirio tŷ neithior*) and the *stafell*, literally 'room', which more specifically denoted the contributions that the bride and groom made towards furnishing their home. Details of 'repairing to the marriage house' are scanty, but when Cerngoch (John Jenkins), who became well known in Cardiganshire as a local poet, was married in 1845 there were fifty people present on 'repairing to the marriage house', and they contributed nineteen cheeses, wheat, barley, and money. On a similar occasion when a South Cardiganshire farmer's daughter was married in 1817 there were received 424 pounds of cheese, three pecks of barley, one peck of flour, five bowls (of unspecified size) of wheat, three bowls of flour, one pound of sugar, and bread. These contributions were made either as repayments of debts owed, or as dues to be repaid by the recipient. The goods received were auctioned and contributed to the sum available to the newly married couple, as did the money raised by selling cakes and drinks to the younger men and women who congregated in the barn when the bidding was held.

Men and women, farmers and cottagers, farm children, servants, labourers, and maids attended and contributed at biddings regardless of whether the young couple were the children of farmers or of cottagers. These people might come from a wide area, Figure 11 shows the addresses of some of those who contributed at a bidding

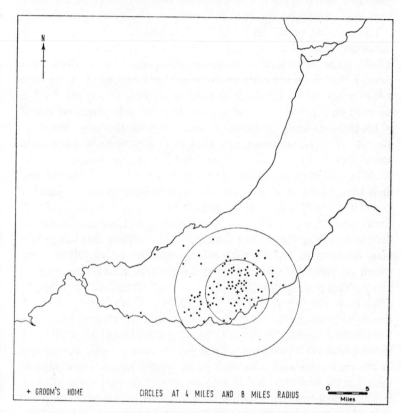

+ GROOM'S HOME   CIRCLES AT 4 MILES AND 8 MILES RADIUS

Fig. 11. Identifiable addresses of contributors to a groom's bidding (1833).

held near Llanwenog in 1833. (The other addresses are no longer identifiable with certainty if at all.) At a bidding held about 1880 on the marriage of a maid employed at a farm on the Bronwydd estate, the bidding was held at the farm-house not at the maid's home, and Sir Marteine Lloyd of Bronwydd was present. The last bidding that I have been able to trace in South Cardiganshire was held in 1887, but biddings continued for some further years in mid

Cardiganshire. At those held within living memory a table was placed near to the entrance to the farm-house. On it stood a bowl in which people placed their cash contributions as they entered the house, and a clerk recorded their contributions. (Alternatively two bowls were used and people placed contributions towards the groom's bidding in the one, towards the bride's bidding in the other. Two clerks recorded the contributions.)

The *neithior,* or bidding, was in effect and among other things a customary form of savings club. A man (or woman) started to contribute sums of money or make contributions in kind during his teens, he 'drew on the club' on marriage, and continued to pay into it for the rest of his life, both in order to repay contributions he had received on his marriage and in order to claim repayment on behalf of his children when they came to marry. David Davies of Penalltygwin in Troedyraur parish, and later of Galltycnydie in Llangynllo parish (where he was farming in 1849), gave 'An Account of the Weddings which I David Davies has been since I were born'; he had then been to 102 biddings and his wife to seventy-one.[12] Another David Davies of Tan y ffynnon, near to Llwynrhydowen in South Cardiganshire, either attended or contributed to 154 biddings during the nineteenth century. Here then was an institution that involved a man for most of his life with a large number of his fellows, and which provided for one of the critical periods in any family cycle, establishing people upon marriage. And as farmers' children had to delay their marriages until there was sufficient provision to establish them, the bidding allowed of earlier marriages than would have been the case otherwise. Further the bidding contributed to integrating individuals into a community for it gave people a particular interest in the marriages and relationships of others, and a knowledge of their family histories, for it was on important occasions in these families' histories that bidding dues were rendered.

By the late nineteenth century the bidding was in decline and the reasons for this are complex. The bidding was then a changing institutionalized procedure within a society that was itself in process of change. One of the relevant general changes may be noted, namely, emigration to the industrial areas of South Wales and elsewhere and a partial replacement of the emigrants by people without any prior connection with the area. The bidding depended on the repayments of debts over a lifetime by people who had contracted them on

12 N.L.W., MS. 9956B.

one occasion and on the making of contributions by those who expected to claim repayment either on their own behalf or on the behalf of people who were their 'near relatives' (in the sense indicated on p. 108). But once continued emigration made it uncertain whether a young man would remain in his native area or emigrate he was in no position to know whether he would be in a position to reclaim against any contribution that he had made, nor would others know whether debts owing to them would be repaid by people who might emigrate. There are indications that emigration was affecting certain of the institutions of South Cardiganshire by the 1870s. Until that decade farm servants were hired at the annual hiring fairs at Newcastle Emlyn, Cardigan, and other centres; during the 1870s it became common practice to hire in advance of the fairs because there was a shortage of servants as the result of emigration. The fair increasingly became a pleasure fair while the hiring of servants was undertaken independently. In the same period it became the practice to send 'industrial schoolboys' into south-west Wales to be employed as farm servants as an insufficient number of local youths was available. These industrial schoolboys had no place in the area as members of kin groups nor was there any certainty that they would remain in the area. Thus the conditions upon which the bidding depended, that there were people who tied themselves for a lifetime expecting that their future would be spent in the same general area, were rendered uncertain and for this reason among others the bidding declined. When this happened there was no longer any institutionalized provision, involving some hundreds of people, for establishing people on marriage and the whole burden fell on the individual families concerned. In the depressed conditions of the closing years of the nineteenth century many farmers felt themselves obliged to delay their children's marriages as they could not spare the live and dead stock that would be required to help set them up on farms of their own. More than one informant has commented on the difficulty that a farmer found himself in when he could scarcely afford to part with his stock but could only keep it by hindering his children's marriages.

As agriculture was less mechanized then than it has subsequently become, and as machines could be borrowed from neighbours, those who intended marriage knew that when they entered into the occupation of a farm there would be relatively little machinery that they would have to acquire immediately. They would have the services of the work group attached to their farm, and the help of those farmers with whom they themselves would constitute a co-operative group. Their major requirements would be livestock and feeding-stuffs, especially hay and corn, and it was for these that they would be chiefly dependent on their parents, and for what was necessary to furnish the dwelling-house.

When a marriage was intended and long before the wedding date was fixed, indeed long before it could be fixed, the young couple and their parents started looking round for a suitable farm. Whether a farm was suitable or not depended on whether the families of the young man and the young woman could afford to stock it without unfairly affecting their capacities for acting similarly when their other children were married. Finding a suitable farm could take a long time, for farm tenancies fell in only once a year, on Michaelmas Day, the twenty-ninth of September, and failure to find a farm in time could mean a year's delay before there was another oppor-tunity. It was at this time that the help of kinsmen was of such importance for then the chance of finding a farm could be linked with the movements of relatives if and when their circumstances made a change of farm desirable.

Arensberg has described how in Ireland the parents of an intend-ing bride and groom engaged an intermediary to help in arranging the details of the 'fortune' that the young woman would bring with her.[13] There is no memory of employing an intermediary in Cardigan-shire in order to discuss what each of the two families could contribute towards establishing a young couple, but a late eighteenth-century account of making marriage arrangements states that the details were settled with 'some responsible persons assisting on either side. . . . This they call *Dyddio*'.[14] The words *dyddia, etidda, notydda* were variously employed in South Cardiganshire until the late nineteenth and early twentieth centuries to denote the occasion when the young man's father visited the young woman's parents in order 'to get understanding' (*i gael dealltwriaeth*) on what could be provided to stock the farm that the young couple would occupy on marriage. *Dyddio* or *dyddia* is to fix a date, and it should be borne in mind that a marriage was arranged as and when a farm became available, as has been mentioned above.

---

[13] C. M. Arensberg, pp. 73–4.
[14] L. Morris, 'Cardigan Weddings', *The Gentleman's Magazine*, vol. lxi, Part 2 (1791), p. 1103.

It has been noted too (p. 45) that though farmers gained their cash incomes chiefly from the sale of store cattle and dairy produce, the annual routine of farm work was dominated by the arable cultivation that was needed to raise bread corn 'for the use of the house' and feeding-stuffs for the animals. Ploughing and sowing were finished by the end of April and the coming of May, and thenceforth work was lighter until the hay harvest started in late June and early July. The tempo of work then slackened again till the corn harvest began in August to be followed by the autumn ploughing and the root harvest. By late October this was over, and with the days drawing in work was less pressing and there was more time for leisure. In mid Wales farm tenancies changed during the spring while maids and servants were hired as from the first of May. May then was a time when work was light, when farm-houses had to be vacated by departing tenants, and when farm servants' annual contracts ended. It was also the month in which most marriages were celebrated, when farm-houses became vacant for incoming couples, when maids and servants were released from their contracts and those who intended to marry were free to do so. There were roughly twice as many marriages during May as during any other month.[15] On the other hand in Troedyraur most marriages were celebrated either in February before the sowing season was under way, or in September and October, which was the time of the year when tenancies fell in (Michaelmas) and the time too when the slack period of winter began.[16]

It has been seen that when sons and daughters worked as members of a family and for the common profit of the family, the marriage of any one concerned the others too. While it concerned the bride and groom most of all, it also concerned their families, their kin, and their neighbours. This received recognition in the use of the term 'to start one's world' (*dechrau byd*) to refer not to birth but to marriage. People would say as they still do that so and so had 'started his world' at such and such a farm. At that time a man's status changed *vis-à-vis* all other members of the community. Until honeymooning became the common practice during the early years of the twentieth century the newly married couple repaired to their new home after the wedding ceremony, accompanied by their relatives and friends. If a bidding was held at all it was held during the

---

[15] A. D. Rees, p. 26.
[16] Information from the General Register Office.

afternoon. In any case friends, relatives, and neighbours were present during the afternoon when they 'went to see' (*mynd i weld*). What they went to see was the '*stafell*' (room), that is the contributions that the bride and groom each made towards furnishing the house. Each had a customary contribution to make: the groom's 'room' consisted of a table, dresser, chairs, and bed; the bride's 'room' of bedding and bed clothes, kitchen and dairy equipment, and an oak chest. People also 'went to see' those gifts already given by relatives and close friends. There was and there remains a dislike of giving money gifts at weddings and a preference for items that can be displayed (such as vases and jugs on a dresser), by which the donor will be remembered, and which the recipient can point out as the gifts of such and such individuals. When people 'went to see' it was normal practice to keep open the settle that was a feature of every kitchen. A settle was divided into two or three compartments, and one of these was used as an airing cupboard. Here were displayed the stockings that the young woman had knitted as part of her contribution to the 'room'. Other neighbours and friends brought tea, sugar, flour, and bread, and though the bidding has ended it remains a matter of honour to make return gifts to the donors on their marriages.

While a marriage was an occasion for sentiment and celebration and congratulations for the young couple and their families, the change of status that it marked had more than a sentimental basis. Henceforth a young man who had been a member of that group of young men that every locality knew became *mishtir* (master) not only to servants employed at his farm but to all those who set out potatoes at the farm and paid their labour debt at the harvest time. The young woman became the *mishtres* (mistress) who saw to providing buttermilk once or twice a week for each family that belonged to the farm's work group, after every churning, and oatmeal on occasion. The couple took over the place that the farm had in a group of farms that co-operated at various times of the year. They became 'farmers'. The fairs that were held at the market towns were frequented by dealers who came to buy casked butter and stock. As the fairs were attended by those who conducted the financial business of the farms, that is the farmers, they constituted what were virtually farmers' gatherings. (Servants were released only for the two or three most important fairs during the year.) A man first attended these farmers' gatherings regularly on taking over the financial business of a farm, which was usually on his marriage.

Rees has described how in mid Wales the way in which a man was referred to and addressed changed on the occasion of his marriage.

A married occupier of a holding is called either by his Christian name and surname together—Evan Evans or William Jones—without any other appellation such as Mr., or else as *Gŵr–man–hyn–a'r–man* (the husband of such and such a farm . . .). An unmarried son is never referred to in either of these two ways, but by his Christian name or one of its diminutives followed by the name of his father's farm. Even though he be an elderly bachelor farming a holding inherited from his parents, he is still *John Tŷ Uchaf*, or *Wil y Wern*.[17]

No such clear-cut practice existed or exists in South Cardiganshire. Strangers who came to farm in the area were always referred to by their surnames with the names of their farms appended, *Jones y Goilan* or *Dafis y Wein*. Those who had grown up together and had been members of the same 'age group' of farmers' sons and farm servants, addressed each other by their Christian names whether they were farmers or cottagers, or the one a farmer and the other a cottager. How people were referred to (as contrasted with how they were addressed) varied with their personal characteristics, many married farmers being known by their Christian names along with the names of their farms. But there was a tendency to refer to the man who conducted the farm's financial business and who dealt with the dealers at the fairs by his surname, with the name of his farm frequently appended. This happened to most farmers on marriage, but to others when the death of the father left one or more children at home. Three brothers who occupied a farm with their two sisters (all middle aged and unmarried) were referred to as Wil, Evan, and Jones. Jones did the dealing, and people had to think for some time before recalling his Christian name. In another case where a brother and sister had inherited a farm and the sister married, her husband (who had joned them at the farm, an unusual occurrence) was known by his Christian name along with the name of the farm even when children had been born to them. He was in no way regarded as an extra hand on the farm but he did not undertake the dealing at markets and fairs. In so far as the way in which a man was referred to did change, it did so when he undertook the farm's financial business, and this generally happened on his marriage, but not invariably.

[17] A. D. Rees, p. 64.

VI

FAMILY AND FARM: (2) AGE AND SUCCESSION

MENTION has been made of different ways in which family arrangements were institutionalized in rural communities in Ireland and Wales. It has been seen that in South Cardiganshire children working on their parents' farms left home on their marriages to 'start their world' on holdings of their own. It has been noted how a childless couple or a couple whose children were too young to work might need to employ servants to work the farm. It has been noted too that when the children were older farm servants were dispensed with as far as possible and the family accumulated the capital which went to establish the children on their own farms when they married. It is now time to discuss the later stages in the farm family's history, and to describe the customary provision that was made both for the declining years of the parents and for the succession to the occupation of the holding. This is best done by considering in the first place the changing composition of the familial core of a farm household in the course of time.

In the folk image of the farm family the children spent their early working years at home with their parents, leaving to work elsewhere only if circumstances rendered it essential. This, as was mentioned above, was the time when the capital was acquired wherewith provision could be made for the future of the children. In the folk image again children remained unmarried while at home, leaving to enter farms of their own on their marriages. One at least of the sons would remain at home during the parents' old age, remaining unmarried during his mother's lifetime and eventually succeeding to the occupation of the holding.

It can be seen from the accompanying table (Table 4) that the folk image corresponded to the actual practice during the opening years of this century in that children did not marry while living in the parental home. It can be appreciated too that the fact that children did not normally marry while living in the parental home affected the

TABLE 4

Troedyraur parish farm households: composition in respect of kin

	Number of households	
	1900	1861
1. Married couple, or parents with young child(ren)	15	11
2. Parents with working child(ren)		
(a) Child(ren) unmarried	16	12
(b) One or more child(ren) married	0	1
3. Widow or widower with		
(a) Unmarried child(ren)	5	6
(b) One or more child(ren) married	0	4
4. Widow or widower without children	1	0
5. Unmarried siblings	6	2
6. Spinster	0	1
7. Bachelor	0	1
	43	38

composition of households; it also resulted in a particular sequence of 'typical' stages of household composition, namely households of parents and children (in the earlier stages of married life), households of a widowed parent and one or more unmarried children (in later stages), and on occasion households of unmarried siblings left in occupation of the holding on the death of the last surviving parent. As we are here concerned with the later stages of this series it may be noted that in four of the five cases in which a widow or a widower comprised a household along with the children (in 1900), either a bachelor son was 'master' to his widowed mother or a spinster daughter was 'mistress' to her widowed father. (In the fifth case a widower without daughters employed a 'housekeeper' to fill the role of 'mistress'.)

While it is true that the folk image of the family and the household corresponded to the existing conditions in Troedyraur parish in 1900 in that children did not marry while living in the parental home, actual practice was in fact more varied than is suggested by the details in one particular year for not all the manifold and untoward events of life could be catered for within the pattern of routine expectations. In one household in Troedyraur parish where the daughter was in 1900 at home with her ageing parents she soon afterwards married the servant and remained in her parents' home.

Again in Troedyraur parish in 1900 households of unmarried siblings were prominent, there being six such households, and these are discussed in greater detail below. But it may be said here that in each of two of these six households, both of them comprised of bachelor brothers, one of the brothers married the maid shortly after 1900. In one case the newly married couple remained at the siblings' home, in the other case they left the siblings' home.

These were the events which were contrary to the common expectation, events which occasioned a great deal of gossip and speculation, and events which demanded unusual personal adjustment on the part of the people who were closely concerned in them in order that they might accommodate themselves to the changes in their circumstances. They were not the only events for which there was no place within the framework of routine provision. Marriage following premarital conception was on occasion catered for by each of the newly married couple remaining in his and her parental home respectively; on other occasions families were divided when both the father and the mother came into occupation of farms, so that in one case the mother farmed the one holding with her bachelor son and the father farmed the other holding with a married son and his wife. People could not always accommodate the unforeseen and unusual events of life in a manner that conformed to their expectations of the usual.

Turning to the details of household composition in 1861, it will be seen that in contrast to 1900 households of unmarried siblings were fewer, but households containing married children were commoner. In all the four cases in which a married child was at home with a widowed parent that child was a daughter. In two cases she was at home with her widowed father, being the 'mistress' of the farm. In the other two cases she was at home with her widowed mother though in general there was room for only one mistress on a farm, for the role of 'mistress' was the highest that was open to women other than in the family as such.

This was a society in which men had various public offices open to them and they had numerous public roles to perform, as deacons and officials in a neighbourhood's chapels, and before the end of the century on school boards and in local government. The fairs that were held regularly in the South Cardiganshire market towns and less frequently in the countryside constituted places where farmers met their friends and acquaintances and relations. But farm women were much more tied to the farm-house and the farmyard. They performed very few public roles. They were never chosen for public office nor publicly expressed a voice on formal occasions in the official affairs of the community. In particular people considered that a woman's place was in the home and the family, and those women who had neither families nor standing in the organization of farm work, the spinsters who earned their living undertaking the incidental tasks of the countryside (see p. 44), were the butts of a good deal of ridicule. Meanwhile bachelors were often enough people of considerable standing for there were public offices open to them though fun was poked at them on occasion.

But with far fewer roles open to her than to a master a mistress did not readily share her position in the farm household. When she did it was with her daughter, and it was rare indeed for her to share her position with a daughter-in-law. Correspondingly a daughter had less fear of being no more than a married servant should she share her mother's household than if she shared her mother-in-law's household. Thus more problems were posed should a son bring in a wife to the parental home while his mother was alive than if a married daughter remained at home, and such an occurrence was extremely rare though it did happen on occasion. (In 1861 the one case of a married child at home with the parents was that of a married son who with his wife and daughter constituted one household with his parents.)

Thus the composition of households was affected by (just as it contributed to) the nature and circumstances of the community of which these households formed a part. The circumstances, however, were not constant but continuously changing in one respect or another. During the last quarter of the nineteenth century improved educational opportunities provided an avenue to a new range of occupations so that there was less call to keep sons at home until they could be found farms of their own. In the same period the outlying portions of estates were sold off and once tenants became free-holders they were in a position to establish some of their children by subdividing their holdings, so that there were fewer children who married while still in the parental home. On the other hand this same change, under somewhat different circumstances, led to an increase in the number of households composed of unmarried siblings, in a manner which is discussed below. Changes in the circumstances affecting household composition resulted again from the large-scale sales of estates which occurred immediately before and immediately after the First World War (which made it more difficult to get tenant farms and thus hampered movement from one farm to another), and there were further changes as a result of the intense agricultural depression of the twenties and the thirties which discouraged farm children from remaining in the farming community.

Relevant to the composition of households and the succession to the occupation of holdings is the consideration that the departure of a farmer's children to establish new households of their own on their marriages affected the household which they left in that the farmer might find himself with a holding that he could not work with family labour. The household's resources were no longer consistent with the farm's requirements. Examples have been cited (in Chapter V) that show how the situation was on occasion resolved in whole or in part by moving to a smaller farm. The following instances were noted: one family moved from a 193-acre farm to a 109-acre farm, another from a farm of 144 acres to one of 96 acres, a third from a 140-acre holding to a 33-acre holding, and a fourth who had farmed two holdings with a combined acreage of 104 acres separated the holdings into one of 80 acres and one of 24 acres, and occupied the smaller holding. Most of these holdings were tenant farms, but one of them was a free holding (of 140 acres), and most exceptionally in the case of freeholders the family left it for a smaller holding when they could no longer work it with family labour.

Just as those who did not move to smaller farms needed to employ servants, so too might many of those who did move, according to family circumstances. What needs to be stressed is that when a farmer moved in this way he took his unmarried children with him (as in example 3, p. 111) so that when one of them eventually succeeded him it was not necessarily a case of succession to the occupation of a farm that had long been in the family's hands. Indeed in some cases a son succeeded when the family had been in occupation for only a few months.

In the division of family work on a farm, which has been described above, the father was responsible for the management of the farm and for transacting its business with the dealers, while the mother was responsible for running the farm-house and for the work of the dairy. They retained these duties for as long as they lived, provided that their health allowed. Just as the farmer managed the farm when he had all his sons and daughters at home so he continued to manage it when he had only one son left at home and the relationship between them might be none too cordial. In Ireland, sons of whatever age, at home and unmarried before the marriage of the heir, were 'boys', in South Cardiganshire a son left at home with his parents was referred to among those who knew him by his Christian name with the name of the farm appended as, for instance, Dai Cartrefi or Twm Pencwaler, but to a stranger he would be described as 'the son of such and such a farm', *mab* Penlasallt, *mab* Alltwen, etc. His status in the working arrangements of the farm was that of a servant, but his status in the family that occupied the farm was that of a son, and people were well aware of both these factors in the position of the child who had stayed at home on his parents' farm.

This practice whereby one or more children remained at home with the parents ensured that there was care for the elderly during their declining years. It can still be heard said of a bachelor son in his forties or fifties that 'he is very good to his mother' or 'to his parents' as the case may be. During the later years of their parents' lives the unmarried children were under an obligation to care for them, and this obligation they owed not only to their parents but to those siblings who had already left the parental home. It was generally understood that those who remained at the parental home would care for the parents in their old age, for by then those siblings who had married would have families of their own to care for.

It was extremely rare, though not unknown, for the parents to retire from the farm so that a son could succeed them. One example has been cited (p. 113) where two holdings, each with its dwelling-house, were run as one unit, the family occupying one house and the labourer the other. On the son's marriage the parents moved into the house until then occupied by the labourer, and the son remained in the house previously occupied by his parents and himself and took over the management of the holding. In another case the parents retired to a cottage in a village a few hundred yards from their ninety-acre farm on handing over to their son when he married and it was considered that the parents had come down in the world. On a farm one's workplace is also one's home. There can be no retirement without a place to retire to and financial provision for retirement. While circumstances did not allow this retirement was extremely rare; once circumstances did allow of retirement (after the Second World War) farmers showed no particular unwillingness to retire. Indeed, easily accessible farms occupied by elderly farmers are now the places where farmers who have retired to village houses congregate to chat and discuss the affairs of the day. Yet financial considerations alone do not account for the way in which farmers remained on their farms and in control of them in the past. With no better financial resources Irish farmers withdrew to the 'west room' as Arensberg describes it: there was no corresponding customary provision in Wales.

So far I have spoken as if all the children who left the parental home were in fact established on farms of their own. While this was the ideal it was commonly recognized and accepted that it could not always be realized. There was only a certain number of farms available and family resources were limited. Of the 155 children who survived to adulthood of parents who held thirty-three farms in Troedyraur in 1900, ninety-six were established as farmers and farm women, while fifty-nine left the farming community and do not concern us here.

Bearing these considerations in mind it is possible to turn to the details of sixty cases of succession to the occupation of farms in the Troedyraur area in the late nineteenth and early twentieth centuries. These do not include every case where farms passed from the occupation of one person to another. As is shown by examples already cited (in Chapter V) the occupation of farms on occasion changed as the result of bankruptcies and again when the sitting tenants were obliged to leave when the farms were purchased over their heads. Farms might thus change hands under circumstances which left the occupier with no choice as to his successor, and these cases are excluded from the accompanying table, as is the succession to farms which remained in the occupation of the gentry. The following cases cover succession to the occupation of thirty-eight of the farms in the area and thus constitute a fairly comprehensive account of changes of occupation over a limited period of time.

Before commenting on what is revealed in this table it will be as well to illustrate the complexity of the ways in which certain instances of succession occurred, for the details in the table do not reveal all that is involved. In particular they do not reveal variations in the manner in which one child rather than another succeeded to the occupation of the parental holding. These can be indicated by instancing cases of succession by an elder rather than a younger brother, and of succession by a daughter when there was also a

TABLE 5

Succession to the occupation of the parental holding

	Freeholder	Tenant
An only child, son	0	1
An only child, daughter	0	1
Daughter, where there were sons	1	0
Daughter, where there were no sons	1	0
Only son where there were daughters	1	1
Elder (or eldest) son	4	11
Younger (or youngest son)	5	18
Other son	2	5
Brother(s) and/or sister(s) jointly	6	2
	20	39
None of the children	0	1
	20	40

son. The elder son (A) of the occupier (B) of a 140-acre farm entered a seventeen-acre holding when he (A) married. This he cultivated and also worked as one of the auxiliaries of the agricultural system (as described above, p. 43). His younger brother remained at home with his parents, but left home when his mother died and his father remarried. When the father's and the stepmother's health failed the elder son (A) left his seventeen-acre holding and returned to the parental farm to which he eventually succeeded. By then he (A) had already established his only son, so that on his (A's) death it was his daughter who succeeded to the tenure of the farm. It can be seen that a simple notion of the children leaving home one by one so that the remaining child succeeded hides the variations in how families catered for differences in their respective sizes, compositions, and fortunes.

In the parish of Troedyraur in 1901 there were forty-six holdings of thirty acres or over (that is, large enough to support husband and wife without the need to undertake other work regularly). Three of these were occupied by gentry. Of the remaining forty-three thirty-three were occupied by tenants and ten by owner occupiers. Though the freeholders were in a different position from the tenants few of them had considerable property to dispose of and it is doubtful whether the material standard of most freeholders was superior to that of the tenantry, if only because those who had purchased farms lacked the working capital that was necessary if they were to capitalize on their ownership. Half of the freeholders had only bought their properties late in the nineteenth century when outlying portions of estates were being sold off, and the purchases were frequently on mortgage. (The sources of mortgages that I have been able to ascertain were certain substantial farmers, a number of auctioneers, and some of the general merchants who traded in agricultural supplies as well as in household goods.) When farms were offered for sale and purchased by tenants, the reasons usually advanced were less any economic advantage that accrued as that ownership conferred security of tenure. Nevertheless when the question of succession to farms arose, tenants and freeholders were differently placed in that succession to occupation of freehold property involved its ownership as well as its occupation.

One of the marked contrasts between freeholder and tenant families in respect of the succession to the occupation of farms lies in the number of cases in which siblings jointly succeeded to the occupation of the holding. In the case of freeholder families siblings succeeded jointly in almost a third of all cases. In the case of tenant families joint succession occurred in but one instance in twenty of all cases. It is proposed to discuss this difference below and to examine immediately cases of succession by the younger (or youngest) son as contrasted with succession by the elder (or eldest) son, once cases of joint succession have been excluded.

It can be seen from Table 5 that the elder (or eldest) son succeeded to the occupation of the parental holding with much the same frequency, between a quarter and a third of the instances, in the case of freeholders and tenants alike (once cases of joint succession have been excluded). Respecting succession by the younger (or youngest) son, in the case of freeholders this occurred in a third of the instances (joint succession by siblings again excluded) while in the case of tenant families it occurred in one-half of the instances. That is whereas among tenant families the younger son was left behind on the parental farm on the marriages of his siblings (eventually succeeding to the occupation of the holding), among freeholder families a number of children were liable to remain at home jointly to succeed to the occupation of the holding.

Two matters may be noted. First, when the children of freeholders jointly succeeded to the occupation of a farm it did not necessarily mean that they were joint heirs, as is discussed below. Secondly there were problems of inheritance involved when the children of a freeholder left their parents' holding, while this was not the case among tenant families. Freeholders' children who entered tenant farms on their marriages might well remain tenants while their unmarried siblings became heirs to their parents' farm. Alternatively other, and what was adjudged roughly equivalent provision (such as education leading to a profession, or marriage to a freeholder's heiress), was necessary for some of the children in order to leave at home one child to inherit on his parents' death. One outcome of these complications involved in the succession to a freehold farm as compared with succession to a tenant farm was that continuous succession of parent by child over many generations at the one farm was just as likely in the case of tenant farms as in the case of freeholdings.

When in the earlier years of the nineteenth century a marriage was intended between children of substantial freeholders, the fathers of the intending bride and groom agreed to settle property upon them.[1] This practice seems to have continued in North Cardiganshire after it had disappeared in South Cardiganshire. Farmers were unwilling to let any property pass out of their hands and even when in the late nineteenth century property became subject to death duties farmers were most unready to see it transferred even to their own children. Property was settled only by will. Farmers were, however, extremely loath to make wills. There was a popular notion that to make a will hastened a man's end, and it was rare for anyone to make his will until and unless he was well advanced in years or suffering from serious illness.

If a substantial freeholder made his will before his children

---

[1] There follows a quotation from a marriage settlement made in the 1820s in South Cardiganshire on the occasion of the marriage of the children of two substantial freeholders. The father of the young man undertook to settle a 60-acre farm on his son, while the father of the young woman made a settlement from which the following is quoted. (All initials have been changed.)

'As a Marriage is shortly to be Solemnized Between A. B. (groom) of —— and C. D. (bride) of —— E. D. Father of the said C. D. Doth Promise and Ingage to give to his Daughter AS a Dowry the sum of five Hundred Pounds in worth and Money but the money not to be paid or any Interest thereon till after the death of F. B. Father of the aforesaid A. B. Nor is the money to be paid at all should the said C. D. Die Without Issue.

'To the decease of the aforementioned F. B.—and further should the said C. D. depart the world within the space of three years leaving no child nor children after her, Her Chest and Cloathes together with her Feather Bed are to be returned All this is promised by E. D. in consideration of what F. B. has engaged to settle on the young couple.'

It will be noted that the cash settlement is dependent on the birth of a child, and that the items of property noted in the second paragraph were parts of the woman's 'room' (stafell).

Source: unpublished manuscript in private possession.

married he generally favoured his eldest son, except when he found most or all of his children still at home with him when he came to make his will in his old age. If he devised the farm intact to his eldest son then he divided his personal property between the other children. (Provision was made for the farm to pass to the second son should the eldest die unmarried, and to the other sons in turn should their seniors die unmarried. Should all the sons die then the farm passed to the daughters jointly.) But if the eldest son had married when the father made his will the farm would not then be devised to him for it was considered that he had received his share of the patrimony when he married. The same applied to any other children who had married, and it was then the practice either to devise the farm to the children left at home unmarried, jointly, or if the property was left to one child rather than another, it was generally with the proviso that a home should be provided for the unmarried siblings as long as that was required. It is these practices which explain one of the most striking differences mentioned above between the succession to the occupation of freehold and tenant farms respectively as revealed in Table 5, namely the disparity in the number of cases in which brother(s) and sister(s) succeeded jointly to the occupation of farms. And when unmarried siblings found themselves constituting one household, brother(s) were obliged to their sister(s) as sons were to their mothers, and each was expected to remain unmarried unless the one or the other married away from the farm.

It is these practices which explain the composition of four of the ten freehold farm households (other than those of the gentry) in the parish of Troedyraur in 1900:

		Age
1	Bachelor brother	51
	Bachelor brother	41
	Bachelor brother	40
	Single sister	43
	Single sister	42
2	Bachelor brother	60
	Bachelor brother	56
	Bachelor brother	47
	Single sister	44
	Single sister	40
3	Bachelor brother	37
	Single sister	41
	Single sister	40
4	Bachelor brother	65
	Bachelor brother	60

It has been seen that in South Cardiganshire the family constituted an economic group during the years when the children worked at home and the family accumulated capital. Under circumstances in which there was no generally accepted practice that there should be one chief heir there was always the risk of difficulties when the time came to share out what wealth had been acquired. Disputes over wills and inheritance were a most fruitful source of dissension in the community. Such dissensions included both those between members of the same family and kin, and dissensions between families not otherwise related when property had passed from one family to another via the widow of a man who died intestate.

In the case of tenant families there was no real property to hold the household together in such a way as was to be found in freeholder families, but as the same notion of the right of each child to provision from the patrimony prevailed, households comparable to those noted above were formed when unmarried siblings found themselves in occupation on their parents' death, with brothers again obliged to their sisters as sons were to their mothers. As has been noted six of the forty-three farm households in the parish of Troedyraur, excluding those of the gentry, each consisted of unmarried siblings in joint occupation in the year 1900.

Arensberg noted a case of the public disapprobation of elderly siblings (two bachelor brothers) who had succeeded to a farm and remained on it without bringing in a young relative who would marry and 'keep the name on the land'.[2] Far from there being disapprobation in South Cardiganshire of elderly siblings who by remaining in occupation of a farm prevented the ingress of a young relative who would marry and procreate children to 'keep the name on the land', such households were commonplace and generally accepted in the community, while some farms occupied by elderly siblings became popular meeting-places for others whose family circumstances kept them single.

The comparison made above between succession to the occupation of tenant farms on the one hand and freeholdings on the other connects joint succession primarily with freeholdings. But it remains open to question whether it was a continuation of what had long been characteristic of the society that a considerable proportion of farm households should be comprised of unmarried siblings, or

[2] C. M. Arensberg, pp. 30–1.

whether this was specifically a feature of the late nineteenth century. At that time certain estates (in particular the Gogerddan estate) were selling off properties in South Cardiganshire and a number of erstwhile tenants on purchasing their farms found themselves during the same period and for the first time faced with the problem of allocating real property among the children. It is notable that in 1861 there were only two farm households consisting of unmarried siblings, the one being a freeholding and the other a tenant farm. And the situation at the end of the century was unusual too in another respect in that some farmers who had recently purchased farms that were large by local standards could and did provide for more than one child by subdividing their farms, a procedure that could not be repeated in later generations without excessive sub-division of holdings. It may well be that the composition of many households at the end of the century resulted from unusual conditions which were attendant on the major changes in land ownership that were then occurring.

The details of the succession of parents by their children which are given in Table 5 make it clear that there was no certainty which child (or children) would succeed to the parental holding and in this connection it is worth noting the way in which the word *etifedd* (colloquially *'tifedd*), literally 'heir', was used in Cardiganshire. In South Cardiganshire the word was only used of an only son (and perhaps only if he was an only child). But in mid Cardiganshire *etifedd* was used to denote not the man who would succeed but he who had succeeded to a property. It was only used in reference to a freeholder who had inherited from his parents. Such a man was known as and referred to as *etifedd* with the name of the farm appended. Thus *etifedd* Rhoslas was not a son who would inherit Rhoslas but the owner, married and with children as the case might be, who had inherited from his parents. A certain dwelling house in mid Cardiganshire is still referred to as *tŷ'r etifedd* (the heir's house) not to indicate that an heir would inherit it, but that it was owned by a person who had inherited it. Such *etifeddion* (heirs) had a particular status. Being freeholders they could carry guns and fishing rods on their own land, while the landlords retained the game rights on tenant farms. Until the parish church of Llanrhystud was rebuilt during the mid nineteenth century it contained pews reserved for these freeholders.[3]

[3] D. Evans, *Adgofion* (Lampeter, 1904), p. 7.

While recognizing that there was no certainty which child would succeed to the occupation of a farm (whether as freeholder or tenant) in practice it was more common for the younger or youngest son to succeed than a daughter where there were both sons and daughters. Among freeholders succession by a younger (or youngest son) was roughly as common as joint succession by a number of siblings, while among tenants it occurred much more frequently. Among freeholders and tenants alike cases of succession by a younger or youngest son were approximately as common as the total number of cases in which all other sons succeeded.

If one son was to remain at home unmarried during his mother's lifetime there was an obvious biological advantage that it should be the youngest son, for when he found himself free to marry he could do so at a younger age than could an elder son if it was he who had remained at home. It is easy to believe too that the youngest son would naturally tend to be the one left behind when his elder brothers married. This is how Frazer interpreted the practice of ultimogeniture, as a 'natural course owing to the fact that the elder sons usually separate from the father and from their own households while the younger, or youngest "never severs from the father's root"'.[4] But if these were the only factors involved it would be difficult to explain what W. M. Williams found to be the case in a part of north-west England. In the region he described there was a family system different from the one delineated here in that the families were primarily freeholders and not tenants but in other ways comparable, and in which 'one son remained at home unmarried' on the parental holding.[5] There was no certainty which of the children would succeed, but it was succession by the eldest son to the parental farm that was most common, being as common as all other types of succession put together.[6]

A distinct tendency to succession to the parental holding by the younger (or youngest) son was found by Rees in that part of mid Wales which he described, adding the suggestion that it may derive from the character of medieval inheritance codes in Wales, according to which each child was entitled to a share of the inheritance but it was the youngest son who succeeded to the parental holding.[7]

[4] J. G. Frazer, *Folklore in the Old Testament*, vol. i (London, 1918), pp. 438–9.
[5] W. M. Williams, *The Sociology of an English Village: Gosforth* (London, 1956), p. 46.
[6] Ibid., p. 49.
[7] A. D. Rees, pp. 71–2.

There is, however, an alternative or an additional consideration, namely that subdivision of the patrimony between all the children is easier in the case of tenants than of freeholders, the youngest son remaining to succeed to the occupation of the parental holding.

At the same time this cannot be considered conclusive if only because Fox has shown that on Tory Island (off the coast of Donegal) freehold ownership is to be found in conjunction with the practice not of devising land to one chief heir, but of establishing children by making provision of some freehold land for each couple on their marriage.[8]

In South Cardiganshire there certainly was a feeling that all the children should receive a share of the patrimony and while in fact it was the youngest (or younger) son who most commonly succeeded to the occupation of the parental holding, there was also a feeling that the eldest son owed his parents a particular responsibility. It was he who filled the office of chief servant where the children did the farm's work without the help of servants. And when there were several brothers at home it was the eldest son who familiarized himself with all the work of the farm while the others tended to limit their interests to their own specific tasks. The eldest son's feelings of responsibility to his parents on occasion led him to remain at home while his younger brothers left on their marriages. In each of four adjoining farms in the Teifi valley (though not in the Troedyraur district) it was the eldest son who stayed at home, the father being alive in each case. In each case too it was the eldest son who eventually succeeded to the occupation of the farm.

It has been noted that the great majority of farmers were tenants. In reference to their children it could be said that though the son who remained at home on the parental farm could expect to succeed to its occupation, it does not appear that he was regarded as particularly fortunate but as rather less fortunate than his siblings. There was a widespread opinion that the best part of the patrimony went to provide for those who were established on their own farms, and every one was well aware that a long bachelorhood might await the son who remained at home. His position frequently involved him in a long courtship. Ten or fifteen years was not unknown, to a time when his sweetheart was past the age of childbearing. Courtships were broken too when the woman anticipated that the man

[8] J. R. Fox, 'Kinship and Land Tenure on Tory Island', *Ulster Folk Life*, vol. xii (1966), pp. 1–17.

might not be free to marry for many years. It was also the situation of many who fathered illegitimate children, though not the only such situation. In fact the customary family system put one (or more) of the children in a dilemma.

A son who remained with his parents on their farm was faced with a double pull. On the one hand he desired to marry. On the other hand he had a customary obligation to stay with his parents and care for them, remaining unmarried during his mother's lifetime. Now as in the past there is general approbation of those who respect this customary obligation. Their situation is understood. When they are courting there is a great deal of speculation about what they will do; it is said that their parents have expected to keep them and they are commonly described as being good to their parents when they act as is customarily expected of them. But there was and remains a structural 'built in' point of tension in the family system. The son remaining at home could only care for his parents in the way customarily expected of him by forgoing the chance of marriage for an indeterminate period; he could marry only by violating the obligations he customarily had towards his parents and his siblings. Whether he took the one course or the other depended in the past as now on the personal characteristics, temperament, and sense of family loyalty of those who found themselves in this dilemma.

Generally things passed amicably but where individuals were of a querulous disposition this is where quarrels broke out. When a son refused his obligations his siblings were liable to disown him, and these were the bitterest quarrels the community knew. Most quarrels were soon patched up or glossed over but these quarrels could last a lifetime. Nor was there any certainty how they would be expressed. I frequently heard the saying *'Rhaid mynd a'r gynnen i'r capel'* ('The quarrel must be taken to the chapel') to refer to cases where brothers took opposite sides on chapel affairs because of a family quarrel, or where one brother removed his membership from the chapel attended by his siblings, with whom he had quarrelled.

It has been noted that when siblings jointly succeeded to a holding a brother and sister stood in the same relationship as a son and his mother. Both were entitled to a share of the patrimony according to popular ideas of fairness, in virtue of the notion that there should be a share of the inheritance for each child, and that by maintaining the household as an earning unit there should be provision for each.

But as there was room for only one mistress on a farm, the brother was customarily obliged to remain on the farm and unmarried unless his sister married. (And similarly a sister was obliged to remain on the farm unless her brother married: the only publicly acceptable solutions were that they should accept the situation or that both of them should marry, one of them leaving the farm.)[9] The brother was then in the same position as was a son during his mother's lifetime. He generally accepted the situation. But one quarrel that lasted a lifetime arose when a brother and sister succeeded to the occupation of the holding when their widowed mother died. The brother who was the legal tenant turned out his sister and so deprived her of the provision to which she was popularly considered entitled. There was a conflict between the brother's legal right and what was customarily expected of him. Public opinion condemned him as his siblings did.

[9] This situation is portrayed by Moelona (Elizabeth Mary Jones) in *Ffynnonloew* (Llandysul, 1939).

VII

KINSHIP

THE customary sequence of events within the farm family as described above was one which normally led to at least one child remaining at home with the parents while the other children dispersed to their own homes when they married. But when children thus set up their own homes they do not cease to be members of their natal families, they remain members of their families of origin or of orientation as they have been called, while they establish their own families of marriage or of procreation.[1] The family need not then be a household unit, and here the unit of parents and children living under the same roof will be referred to as the 'nuclear family', while 'elementary family' will denote parents and children whether they are living together or not.

Children then set up their own homes on marriage, but if several of the children were at home and unmarried on the death of their parents they might well remain a household of unmarried siblings. In any locality there were numbers of households linked by bonds of kinship and of affinity. Such households were variously constituted, some being the homes of nuclear families, some of incomplete families having lost one or other parent, others of groups of siblings, and some of people related in other ways. It can be seen from Table 6 that in 1861 in Troedyraur parish households of nuclear families and of married couples between them amounted to less than half the total number of all households and to half the total number of farm households, while in 1900 they accounted for slightly less than two-thirds of the farm households. In this the situation was not unlike that in certain rural areas of Northern Ireland where Mogey found that in the Braid valley of County Antrim 39 per cent of households had neither married couple nor nuclear family in them, and that in the case of Hilltown in County Down the corresponding figure was 52 per cent.[2] Thus while

[1] e.g. M. Young and P. Willmot, *Family and Kinship in East London* (London, 1957), p. 202.

[2] J. M. Mogey, *Rural Life in Northern Ireland* (Oxford, 1947), op. cit., pp. 127, 92, quoted in L. Lancaster, 'Some Conceptual Problems in the Study of Family and Kin Ties in the British Isles', *British Journal of Sociology*, vol. xii, No. 4 (1961), pp. 317–33.

it was households therefore rather than nuclear families that were
linked together by the ties of kinship these individual households
frequently contained people who were members of the one elemen-

TABLE 6

Composition of households in respect of kin: parish of Troedyraur

	1861				1900	
	All households		Farm households		Farm households	
	No.	Percentage of all households	No.	Percentage of farm households	No.	Percentage of farm households
Total number	233	100	38	100	43	100
Husband and wife	32	14	1	2	2	5
Nuclear family	79	33	18	48	25	58
Widow(er) or unmarried person	27	13	2	4	1	2
Otherwise constituted	95	40	17	46	15	35

tary family. Thus families of origin and families of marriage overlapped and spanned three generations, and in doing so constituted
the nuclei of groups of households which were associated with one
another and involved in one another's affairs by the ties of kinship
and affinity.

The recorders' notebooks of the census of 1861 show that 54 per
cent of the residents of Troedyraur parish had been born in the
parish, the birthplaces of a further 27 per cent were in adjoining
parishes, and of a further 9 per cent in neighbouring (but not adjoining) parishes, a total of 90 per cent.[3] It is consistent with the movement from farm to farm that has been described that a little over
half its population should have been born within Troedyraur parish
and most of the remainder in the surrounding area. By 1861 the
population was already declining. The population of the agricultural
areas of South Cardiganshire generally reached its maximum during
the 1840s excepting only where there were woollen mills. From that
time on migration more than balanced the natural increase. On the
other hand immigration into the rural area of South Cardiganshire
was on a very limited scale until the Second World War and the
years following so that the proportion of the population which had
been born in the area remained high, and though I have no complete figures for Troedyraur parish in 1901, those that I have are
virtually identical with those of forty years earlier.

Thus individuals grew up knowing other people as the fathers,
mothers, brothers, and sisters of their school friends and neigh-

[3] Sources in the Public Record Office.

bours. In later life they would come to know youngsters as the
children and grandchildren, cousins, nephews, and nieces of those
with whom they grew up together. That is they grew up with a
knowledge of 'who people were' in a locality rather than needed to
acquire such knowledge *in toto* by deliberate effort as they would
were they to emigrate to a strange community. And the knowledge
acquired in this way was knowledge of people in a community
rather than a systematic genealogical knowledge of kin as such.

The small scale of the society was a general condition of social
life and within a locality people were inescapably concerned in one
another's affairs. This is reflected in what one is certain to be told
sooner or later at the present day, that all the people of such and
such a locality are related to one another 'if you go back far enough'.
If one then points out a number of people without relatives in the
district the reply is a tautological definition, that the 'people of the
locality' or the 'people of the place' are those who are related to one
another, and people without relatives are discounted as not being
'from the place' though they may have lived there for decades, and
perhaps even if they have been born there should it be the case that
their parents originally came to the district as strangers. The aim of
this chapter will be the limited one of considering certain of the
interconnections between the ties of kinship and affinity on the one
hand and those of locality and occupation on the other.

Talk of relatives and of relationships is pervasive, and not only
about the speakers' own kinsmen but about other people's as well.
Conversations are shot through and through with references to the
relationships of people who are the subject of discussion. There is
a general expectation that people will know the main facts about
other people's kin connections while their remoter connections are
readily the subject of inquiry. A good deal of talk turns on how
kinsmen behave towards each other and how what befalls a person
will affect his kinsmen and their relations with one another. The
ordinary turns of fortune, towards and untowards events, marriages,
the birth of children, movements from one locality to another are
seen not as things that affect only the people immediately concerned
but according to their likely influence on the relationship of a body
of people comprised of kin and affines.

But remarkably widespread as is the interest in kin it is also true
that knowledge about kin, including one's own, varies from person
to person in considerable measure. While some individuals have

extensive knowledge of people's kin relations others are little interested. It is said of some that they hardly know who their own first cousins are. Thus while interest is widespread it is neither universal nor indispensable and it does not appear that this is something new. More than once elderly people who had grown up during the closing years of the nineteenth century mentioned kinsmen to me adding that only recently had they found out that they were related. 'If there was no intercourse between them', it was said, 'they became strangers to one another', and the more distant genealogical links were overlooked.

But while this knowledge is and has been incomplete and has varied from one individual to another according to personal interests, it remains the case that people generally relate individuals to one another in terms of their kin affiliations. One has the impression that there is a community of people who can be discussed because their relationships are known and that because their relationships are known they are recurrently brought to mind. This can apply only to those who have kin in the locality and while today there are numerous immigrants without kin connections such people were exceptional during the period with which this study is chiefly concerned. Conversation readily gives the impression of an interest in and a concern with kinship relationships, and sixty and more years ago the interest and concern were certainly no less than they are today.

In pursuit of the aim of this chapter we shall first consider some of the terms which people use in the local idiom in order to refer to and to distinguish between relationships. It will then be practicable to see what people have expected of kinsmen in this society, and to see further in what way kinsmen have interacted with one another or formed groups (if any) of kin, recognizing themselves as such and acknowledged by others to be such.

The word used to denote a kinsman is *perthynas* and *perthynas* means both a 'relative' and a 'relationship'. There is no one word which unambiguously means a kinsman and nothing else. *Perthynas* is used inconsistently sometimes so as to include affines who are recognized as relatives, sometimes to refer only to consanguineous kin. When speaking about a man's affines it is commonly said that there is a 'relationship' between them and him, and then this is qualified by adding 'but through marriage'. On occasion however one hears it said that an affine is not 'related' to the person concerned, but is for example, a spouse of that person's kinsman. The confusion reflects an awareness on the one hand of social relationships between affines and on the other that affines do not normally share common descent.

Those with whom a man shares descent are referred to collectively as his 'people' (*pobol*). One commonly hears it said of certain individuals that they are 'the same people', or an individual will be reminded that certain others are 'his people'. The word *pobol* ('people') can be used for any collective, as for instance *pobol y lle* ('the people of the place', i.e. local people) or *pobol y capel* ('the people of the chapel', i.e. those who regularly attend a nonconformist chapel). Whether a particular application of the word is to a person's neighbours or his kinsmen will be understood from the context, but when reference is to his kin it refers to consanguineous kin and not to affines.

Two other words (*tylwyth* and *cether*) are used in the same manner and are not used of any collectives other than kinsmen except in an obviously figurative sense upon occasion.[4] Despite this a man's kin is generally referred to by the word *pobol*, which bears the most general meaning.

The word for 'family' (*teulu*) may mean children, nuclear family, elementary family, or a body of relatives. It is common practice to say that such and such a person has no 'family' (*teulu*) when he is childless, or to ask a person if he has any 'family' (*teulu*) when one is asking if he has any children. At the same time parents and children are referred to collectively as the 'family' of such and such a house (e.g. *teulu'r Plas* or *teulu'r Gilfach*) while *hen deulu* ('old family') is used to refer to a body of relatives long established in a locality. The more prominent of such 'families', whether prominent for their possessions or public activities or for their numbers, are commonly referred to in one or more of a number of ways. When some members of an 'old family' have long been in occupation of a particular property whether a farm or a dwelling-house, they are referred to as the 'family' of that house; if the house has passed from father to son for several generations then they may be referred to by the surname of the members occupying the property, or by the surname in combination with the name of the house. More commonly however relatives are referred to by an appropriate surname

---

[4] The colloquial form is *cether* which does not have the pejorative associations attributed to *cethern* in *Geiriadur Prifysgol Cymru*.

(as the *Dafisied, Tomosied, Jencinsod, Bowenied,* etc., i.e. the Davieses, Thomases, Jenkinses, Bowens, etc.), and the natal surname group of spouses separately indicated in each individual case.

Outside the elementary family relatives are differentiated and referred to in two main ways. They are differentiated in a manner that is clear but structurally unimportant into the 'two sides of the family'. Certain relatives are either 'on the father's side' or 'on the mother's side' but this distinction is in itself simply schematic. Kinship is bilateral and whether specific relations are more or less important depends not on which side of the family they lie but on other considerations which may be personal and individual.

In the second place relatives are classified according to genealogical distance from the person concerned into 'near' and 'distant' relatives. This distinction is vague though not meaningless. There is no certainty where the boundary between 'near relatives' and 'distant relatives' in this sense is to be placed, excepting that first cousins are considered 'near relatives' in respect of genealogical distance from the person concerned regardless of whether actual relationships be close or not. Nor is there any certainty concerning who is to be included among 'distant relatives' and who excluded altogether. Sometimes people include only those with whom they can trace relationships by more or less surely known links; at other times they include people whom they have only heard mentioned to be their kinsmen. And as far as can be ascertained these practices were much the same at the turn of the century as they are today. Accidents of knowledge, propinquity of residence, and personal considerations have all had a part to play in deciding who is mentioned as a 'distant relative' and who is not. Such kinship terms do not discriminate carefully between people, and one can see the same tendency in the occasional use of kinship terms to refer to people who are not kin at all. Thus children are encouraged to address and to refer to their parents' close friends as *wncwl* (uncle) and *anti* (aunt).

These usages are so obvious to those familiar with them that they appear self-evident. Thus it is worth noting that variations in the use of kinship terms are possible within the compass of the Welsh-speaking rural areas. The standard Welsh word for 'uncle' is *ewythr* but colloquially *wncwl, nwncwl, yncl* (all being derivatives of 'uncle') are widely used. *Dodo* or *bodo* is used in some areas either for a grandmother or an aunt. The standard Welsh word for 'aunt' is *modryb,* but *anti* and *nanti* (derivatives of 'aunt') are in common use. The following variation in classification and terminology has been used by some people in a part of the county of Merioneth.[5]

In it the Welsh forms have not been displaced by the cymricized forms as one might all too readily assume, but have been supplemented by them. The additional kinship terms thus made available

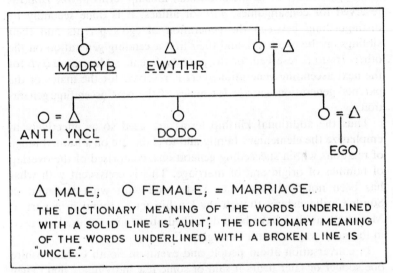

Fig. 12. Variant Kinship Terms

have been used in two ways, to distinguish between consanguines and affines, e.g. between a parent's sister and a parent's brother's wife, and to distinguish between different generations of relatives. These terms are here described in so far as they have been reported. Ego refers to his mother's sister ('aunt' in the standard vocabulary) as *dodo,* but to his mother's brother's wife (also 'aunt' in the standard vocabulary) as *anti*; he refers to his mother's aunt (but not to his own aunt) as *modryb* though the standard meaning of this word is again 'aunt'. He refers to his mother's brother as *yncl,* but to her uncle as *ewythr* despite the fact that in the standard vocabulary it is to ego's uncle that the word *ewythr* applies. Similarly ego refers to his father's uncle as *ewythr,* though in the standard vocabulary this word designates his own uncle.

[5] F. W. Jones, *Godre'r Berwyn* (Cardiff, no date), chapter iv, especially pp. 52–3.

No explicit statement is made about how his father's other relatives are referred to. No information is given to indicate whether there is any further significance to this variant classification and terminology. But what is done through this practice is to emphasize the parent's sibling group. This is done first of all by distinguishing between consanguines and affines in the parent's generation, and classifying them apart in so far as the additional kinship terms allow. *Dodo* is reserved for consanguines, *anti* for affines. It is done secondly by distinguishing between the generation of ego's parents and their siblings on the one hand, and their first ascending generation on the other. *Dodo* is reserved for the parents' female siblings, *modryb* for the next ascending generation; *yncl* is reserved for the males of the parents' generation, *ewythr* for males of the next ascending generation.

Thus the additional kinship terms are used so as effectively to emphasize the elementary family and to elaborate on a limited range of relations within succeeding generations, comprised of the overlap of families of origin and of marriage. This is consistent with what has been noted elsewhere in Wales and with what is found in South Cardiganshire, though the kinship terms found in a part of Merioneth as they have been described above are not used, or used in the same way in South Cardiganshire.[6]

In conversation about people and events in South Cardiganshire one sooner or later hears it said of some few individuals that 'They are "near relatives" but they do not *arddel* one another' or that 'They are "near relatives" but they do not *arddel perthynas*', that is they do not *arddel* the relationship between them. *Arddel* means to profess, as for instance to profess religion (*arddel crefydd*), to acknowledge, and in the context of kinship to behave towards a kinsman in the manner and according to the norms considered fitting. On the other hand to *diarddel* is to take the contrary course; it is to withdraw the hand of fellowship. Not to *arddel* a relationship, or not to *arddel* one another is to behave as if the norms of conduct considered proper between relatives did not apply. It does not mean being on bad terms with a kinsman but to disavow the relationship.

It is the case that the main institutions of the society are such that a man can choose not to 'avow relationship' if he so decides. Kinship considerations do not so dominate that a man in practice has

---

[6] e.g. C. Rosser and C. Harris, *The Family and Social Change* (London, 1965).

no option but to 'avow relationship' and this is certainly not a development of recent years. At the same time people are held by others to be 'near relatives' if the genealogical facts warrant it whether they 'avow' one another or not, and in this too there is nothing new.

Others have differentiated between kinship knowledge on the one hand and kinship recognition on the other.[7] Kinship knowledge refers to a factual knowledge of genealogical relationships without implying any necessary contact or interaction. Kinship recognition on the other hand refers to an active recognition of kinship with mutual contact and participation in each others' affairs.

While it is this mutual involvement which is the meat and marrow of kinship yet in this community kinship knowledge has been of consequence even when it is knowledge of the genealogical connection between individuals who disavow any relationship. This is *in part* because kinship knowledge has been used to provide a means whereby people can be indicated and referred to. It was expected that everyone in a locality should know everyone else and with something like 90 per cent of the population of a parish born within the area of that parish, of adjoining parishes and of neighbouring (but not adjoining) parishes, kinship knowledge has provided the schema needed for 'placing' people and explaining who they are.

When someone is questioned as to who some third person is the answer whenever possible consists of a statement of the links of kinship between that person and people acquainted to both the questioner and the answerer. To this the occupation of the third person may be appended: 'He's a cousin of Jones Coedycwm, their fathers were two brothers. He farms Groesfach', or 'His wife's related to the people who're farming Frongoch'. Failing this the answer consists of the occupation of the man about whom one inquires, and in default of this, too, where he came from. 'He's a Carmarthenshire man' or 'He's from the top end of the county'. The preferred way of 'placing' people by relating them to their kin is old-established. It occurs consistently in regional literature, for many decades.[8] It is used by elderly people who say that that is how they always heard people described. But while it provides a method of referring to people it has further significance in that

---

[7] Ibid., p. 194.
[8] e.g. in the published work of Moelona (Elizabeth Mary Jones), a native of South Cardiganshire.

stating 'who a person is' is not simply to indicate him but to attribute to him a social identity, as is described below.

In so far as kinship knowledge is used for 'indicating' contemporary members of a community the emphasis is on knowledge of a wide span of relationships which may be only of shallow genealogical depth, sufficient to indicate how people are connected to one another. Consistent with this is an unusual naming practice that was fairly general in the early years of this century, and which refers to what many people were called rather than to what they had been christened. The practice was to refer to someone by his (or her) Christian name, followed by the Christian name of one or other of his parents, and generally by the Christian name of one of his grandparents. This is easily illustrated. In one village it was only with difficulty that people could (in 1960) recall the surname of a man invariably referred to as Dai Mari Dafi, that is, David (the son of) Mary (the daughter of) David. Ifan Dafi Dai, who was Evan (the son of) David (the son of) David provides another example from among many. There does not appear to be any regularity in the way the name of one parent rather than another, and of one grandparent rather than another, were selected. It was very rare for the composite name to include names from more than two ascending generations relative to the person involved though it did occur on occasion. A man who had emigrated to a South Wales mining town explained who he was to an old woman when he visited his birthplace by saying, 'It's no good telling you that I'm David Jones, I'm the son of Peggy Daniel Sally Betsy'.

It may be remarked in passing that in a community on Tory Island which had no resident gentry, and with a system of land holding markedly different from that of South Cardiganshire, Fox connected a comparable naming practice with land ownership and with claims to land.[9] In South Cardiganshire this naming practice lasted longest among landless cottagers. One assumes that the mode practised by the landed gentry had at an earlier period been applied to small owners and occupiers of land.

It has been stated that to explain 'who a person is' by relating him to his kin involves more than simply indicating him. It involves the notion that knowing 'who his people are' is relevant to knowing an individual. To relate someone who is unknown to someone who is

---

[9] J. R. Fox, 'Kinship and Land Tenure on Tory Island', *Ulster Folk Life*, vol. xii (1966), pp. 1–17.

known relates him not to one person but to that person's kin. Unlike situations which bring together people without previous connections, in a locality where a high percentage of the inhabitants are native born, people's 'origins' are known no matter what claims or disclaimers they may make for themselves. It has been commonplace to hear people's characteristics, physical and social, explained by stating that they are attributes of those people's kinsmen. Shared features have been attributed to a common origin in a saying well known in the locality and which applies to kin: *mae natur y cyw yn y cawl*, the nature of the chicken informs the broth.

To know who a person is, then, is not so much or not solely to know him 'as he is in himself' as it were, as to know him as one of a body of kin. The link between kinsmen has been held to be 'in the blood'. Personal characteristics have been ascribed to one's 'blood', the ability to write poetry, desirable and undesirable personal qualities, as for instance good nature and querulousness, industriousness and indigence, achievement of prominence as public figures, or success in a particular occupation or profession. When a boy from a successful sea-going family went to sea people would say, 'He'll soon be a captain, it's in the blood'. On the other hand when a child from an undistinguished family achieved scholastic success people would say that it was surprising because there was 'nothing in' any other member of his kin. And on occasion it might be sufficient to prompt people to suggest that the boy's parentage was open to question. It was known too for a boy who had distinguished himself in an unusual way to have his parentage attributed to some member of a group of kin who were distinguished in the same way.

The nature of a person's kin was part and parcel of his social identity and regardless of whether individuals 'avowed' or didn't 'avow relationship' with their kin, people at large connected them together as sets of kin who between them largely comprised the population of this rural area.

In the course of discussing how cottages and holdings of different sizes fitted into an agricultural system note was made of some labour dues that were owed to farms by cottagers for setting out potatoes, others that were owed by the smaller holdings to the farms for work done when farms brought such implements as ploughs to cultivate the smaller holdings, and yet others owed by cow places, one-horse places, and the smaller farms for a bull's services. It was pointed out that a scale of status was associated with these labour dues, and to this day the status of those who occupied the larger holdings is echoed when people say, 'Of course, we were the children of a large farm' as the Welsh idiom has it.

But a person's standing could not be tied to the size of his farm if only because people moved from farm to farm to suit their families' circumstances, and these movements might be from larger farms to smaller farms as well as from smaller farms to larger farms. The following instances are recapitulated: (1) A, who had been farming two holdings with a combined area of 104 acres, separated them and retained one of the holdings, of 24 acres; (2) B, who had farmed a holding of 140 acres for most of his life, removed to a 33-acre holding; (3) C, who started farming on a 15-acre holding, moved successively to farms of 46 acres, 110 acres, 193 acres, and then 109 acres. A man's standing could hardly depend solely on a factor as variable as the size of the holding he occupied at a particular time, but required what was considered to be a more consistent reference. People found this in the standing of the other members of his kin group.

There follows a number of examples of the kin relations of a number of married couples which will illustrate what people had in mind when they associated a man (or a woman) with the standing of his kinfolk. (The data refer to *circa* 1900 and 'near relatives' is used to refer to genealogical distance from the person concerned and in the most restricted sense in which the term is used in the locality, namely so as to include only ego's spouse and children, ego's siblings and their spouses and children, his parents, his parents' siblings and spouses and their children, his grandparents and grandchildren.)

1. S was a farmer's wife. She was the 'daughter of a large farm', and she had 'near relatives' in sixteen households in the area. Eleven of these were farm households and they included the occupiers of some of the largest farms in the district. In the other five households S was consanguineously related to the wives of an innkeeper, shopkeeper, a blacksmith, and two labourers. S was married to D who was the 'son of a large farm'. He was himself a farmer and he had 'near relatives' in nine households in the area. Seven were farm households and two were labourers' households. Again it was the wives in these two cases who were D's consanguineous relatives.

It was only very rarely that all the related households within the locality were farm households, but there were occasional examples. The practice of joint succession by unmarried adult siblings on the deaths of their parents contributed to this, whereby several children remained at home, sometimes underemployed, instead of taking other employment. Thus all the siblings were 'farmers' rather than that some of them should have left home to become labourers. It was usual too for the sons of substantial farm families who could not get farms of their own to move to other districts rather than become labourers in the locality. For instance S's brother migrated to another county where he spent his working life as a labourer, but returned to live with a sibling when his working days were over. One further development that made it possible for members of some kin groups to settle all their children so that none of them was a labourer in the locality was that while it had long been the case that some individuals became ministers of religion, by the 1870s some farm families educated one or more of their sons for other professions.

2. J, who was a labourer's son, was a farmer and he had 'near relatives' in six households in the area. Two of these were farm households, two were craftsmen's households, and two were labourer's households. J was married to M, who was a farmer's daughter. She had 'near relatives' in seventeen households in the locality; twelve of these were farm households, the others were those of a publican, a shopkeeper, a craftsman, and two labourers.

3. W, the son of a craftsman, was a farmer. He had 'near relatives' in ten households in the area. One of these was a farming household; the others were those of eight craftsmen and one agricultural merchant. He had married M, a farmer's daughter. She had 'near kin' in three households in the district, all of which were farm households.

4. D was a labourer's son but himself a craftsman. He had 'near kin' in eight households in the district. Four of these were craftsmen's households, three were labourer's households, and one was a farm household. D was married to R, who was a labourer's daughter. She had 'near kin' in five households in the vicinity, those of one craftsman and four labourers.

These were the background connections that people had in mind when they referred to someone's 'people'. When the farmer in the last example was mentioned my informant said, 'Of course, they were labourers', it being understood that *they* referred to the other members of his set of kin. Though certain of them were craftsmen yet they were characterized as one body of people. Similarly the word *gwrwas* was used to designate a man who being a farm servant and with connections similar to those mentioned in the last example, had entered into a farm by marrying the farmer's daughter. The word *gwrwas* is comprised of two elements, *gwr*, a married man, and *gwas*, an unmarried servant. The *gwrwas* was a married man and a master but his family connections associated him with the class of servants. The word characterized the man according to his origins and his kin.

The major occupational distinction in these examples is that between farmers and cottagers, whether the latter were labourers or craftsmen. It was the case in practice that there was a great variety in the composition of different people's 'near kin'. It was not the case that there was a clear division between farming kin groups and cottager kin groups. People evaluated each one of them individually, and one repeats, regardless of whether particular individuals 'disavowed' any relationship between them.

The nature of an individual's kin was thus part and parcel of his social identity, but at the same time the sole group from which he could expect a contribution as of right in order to establish him on his marriage was his own nuclear family. It was only from this body that he was entitled to provision, though he would have expectations of aid from any related households of childless couples on the one hand and households of unmarried siblings on the other. But just as kinship terms were vague and imprecise outside the elementary family so a man's expectations of his kin were generalized rather than particular, unspecified rather than specific. Friendliness was expected of kin along with backing and support in the multifarious situations of daily life, and help at times of difficulty and in straitened circumstances though with an appreciation that under conditions of hardship there was a limit to what material aid people could provide.

This generalized nature of kinship obligations and expectations had dual consequences. On the one hand it made it difficult for a person to make specific demands on his kinsmen as contrasted with simply asking for their aid. It left a great deal to individual feeling and to a degree it tended to blur the difference between relations of kinship and of friendship.

On the other hand this was a society wherein various basic institutions were not based on kinship. Tenancy was a matter between landlord and tenant not between kinsmen. Farm co-operative groups were based on the farm and not on the occupier and his kinship links. Farm work groups were constituted irrespective of kinship considerations and without regard to ties of kinship between those who constituted work groups. The generalized character of kinship obligations provided the flexibility which allowed individuals at one and the same time to behave in an acceptable manner towards their relatives and to take their places in institutions which paid no heed to considerations of kinship.

Again the countryside's inability to support its own natural increase in population has required emigration, and the periodic movement of families from one farm to another has taken people beyond the range of practical day-to-day co-operation. Thus the detailed membership of farm co-operative groups and work groups has always been liable to change and this again has been rendered easier by the flexibility which results from general rather than specific kinship obligations and expectations.

It has been noted that the general character of these obligations and expectations has tended to blur the difference between the relations of kinship and those of friendship, and friendliness has certainly been one of the main expectations in the relationship between those who are kin. Kinship and friendship both call for mutual aid in times of difficulty and for social support in the manifold events of daily life. Indeed it is tempting to interpret the association of individuals in the countryside as groupings of friends partly recruited on the basis of kin.

But it would be mistaken to see no difference between the obligations of kinship and those of friendship in this rural society. While the ideology of kinship involves friendliness there can also be strains between kinsmen which lead to relations which are intimate but by no means friendly. And in the situations of greatest need help was rendered by those who were not simply friends but kin.

The most demanding and sustained aid was probably required on behalf of children who had lost either or both parents, or whose parents were unable or hard put to rear them because of straitened circumstances. Under such circumstances children were always cared for by their relatives. Further, while the terms 'near relatives' and 'distant relatives' are imprecise and need refer only to genealogical distance when parents could not rear their children the children were invariably cared for by their 'near relatives' (in the sense indicated on p. 168) even if this involved removing the children to South Cardiganshire from the South Wales industrial area. In the parish of Troedyraur in 1861 one household in seven included children being reared by their 'near relatives'. In twenty-four households children of incomplete families were being reared by their grandparents, and in a further seven households children were being reared by their parents' siblings.

It may be noted too that it was those who were genealogically 'near relatives' who were most closely involved at marriage biddings during the nineteenth century. When the customary bidding was discussed it was pointed out that it might be requested that the repayment of marriage dues be made not to the person to whom they were owed, but to one of his relatives. Examples were cited of a number of bidding letters where this was requested (p. 132). In all the bidding letters that I have examined it was invariably for 'near relatives' (again in the sense indicated on p. 168) that people required the repayment of marriage dues other than those to be repaid directly to themselves. While the terms 'near relatives' and 'distant relatives' have been vague and imprecise in common usage, they have by no means been meaningless.

It was noted at the outset of this chapter that the overlap of families of origin and of marriage immediately associated numbers of households with each other on the basis of the mutual involvement in one another's affairs of those who are one another's kin. Such households are linked by the lineal ties between grandparents, parents, and grandchildren on the one hand, and by the collateral ties between siblings on the other. The strength of the lineal ties between succeeding generations is attested to by the fact that of the 233 households in Troedyraur parish in 1861 twenty-five contained three generations (of kin) and another six households (each containing grandparent[s] and grandchild[ren]) spanned three generations. One may further note that this link between grandparent[s] and grandchild[ren] is again to be seen in the marriage contract previously mentioned (on p. 149 in the relevant footnote). The marriage contract provided that the bride should bring with her a dowry and a contribution to the furnishing of the house. The dowry was not to be paid should she die without issue, while the main items which she contributed to the house furnishings were to be returned should she die within three years 'leaving no child or children after her'.

While the strength of lineal ties is thus indicated, the link in this society between siblings was a very close one. On each farm siblings

grew up as members of one work group. They worked as an economic unit, and they were mutually dependent when the time came
for each to enter a farm of his own. They were mutually associated
by people at large and were collectively known as the 'children' of
such and such a farm (whereat they had grown up together) long
after they had left that farm and entered holdings of their own,
indeed even for the whole of their lives. When parents could not
rear their children if it was not the grandparents who cared for
them it was the parents' siblings. Again it was common for an elderly
bachelor or spinster occupying a farm to have a sibling's child as
mistress or master on the holding, and when such a sibling's child
was a widow or a widower with children, to provide a home for
them on the farm. And quite apart from the direct emotional and
other ties between siblings they were further connected by their
links with their parents for as long as they survived.

There was thus a familial basis for the way in which households
were linked together at any one time, but the relationship between
individual households, and the number of households associated in
particular clusters, were not static and fixed but necessarily changed
in the course of time as children grew up and married and established
households of their own, and too as other households changed or
disappeared with the death of the elderly. At any one time the centre
of any one cluster of households thus linked with one another was
the household of the parents whose children were to be found in the
other households concerned. But with the death of the parents this
changed. On the one hand new households came to central positions
as their members became parents and grandparents in their turn.
On the other, the links between siblings generally survived the
deaths of the parents with the result that numbers of households
remained associated with one another, their members forming
groups of kin in the sense that and in so far as the personal identifica-
tion of individuals with one another and a sense of common identity
go to make groups. As such groups included affines they were not
identical with the named 'families' (the Davieses, Thomases,
Joneses, etc.) that have been referred to. And as kinship is bilateral
individuals were commonly members of more than one group of
kin, a situation made possible by the fact that inclusion in such
groups has been decided not by rule but *de facto*, being influenced
by considerations such as propinquity of residence, personality
characteristics, and individual preference.

While the links between siblings were links forged over long periods between those who had grown up together, the marriages of siblings saw not only the setting up of new households but brought about new patterns of contacts and connections. They established links of affinity between previously unrelated households, links involving the same expectations of aid and friendliness and support as were expected of consanguineous kin. Foremost among the households whose relations were thus affected on and after marriage were those of the parents of either spouse. They shared a common and often competitive interest in the household which contained their children, each concerned to receive its share of attention and of recognition. And, on the birth of children, while each shared a common interest in the grandchildren, those who were affines in one generation became for the grandchildren 'the two sides of the family', so modifying the pattern of connections which could serve in the years to come as the basis of social relationships.

But relations of affinity affected more households than those of parents in law; they affected two groups of households and the links of affinity could be the basis of personal ties which were very close indeed. Nowhere can this be seen more clearly than in the way in which affines might take the place in farm organization which under other circumstances would be filled by other kin, as in cases where a daughter's husband filled the role of *mishtir* for an elderly widow without sons of her own, where a deceased brother's widow became *mishtres* to an elderly bachelor, and where a brother's daughter's husband became *mishtir* to an elderly spinster.

Within each group of households mutually connected by an acceptance of the ties enjoined by the ideology of kinship every individual has his own personal set of relations constituted relative to himself, centred on himself, and thus different for each individual save only in the case of unmarried siblings. To be a member of such a set of relations provided a man with a body of people to whom he could look for aid on a few but important occasions in life in such a farming community as this was. This is best presented by recapitulating some of the material presented in the preceding two chapters.

In those chapters consideration was given to the customary arrangements that were made to provide for the members of a nuclear family, for establishing the children when they were married, caring for the ageing parents, and succeeding to the occupation of the farm. Consideration will now be given to the part that a person's kin had in helping to secure these arrangements. It is convenient to begin by seeing what became of farmers' children, and here details are recapitulated concerning the 155 children whose parents occupied thirty-three farms in the vale of Troedyraur in 1900. Ninety-six of them were eventually established on farms of their own but fifty-nine left the farming community and are not of immediate concern. Of those ninety-six who were established on farms twenty-five were sole successors to the parental holdings. Another twenty were joint successors to six parental holdings. It is with these forty-five children who succeeded to the occupation of their parents' holdings that the last chapter was concerned.

They were a minority of those who became farmers, for other farms were found for the remaining fifty-one children. These children's marriages were linked with obtaining farms (as has been discussed), and finding farms for those about to marry meant finding suitable farms such as could be stocked on behalf of and run by the young couple. Where one of two children of parents who farmed a large holding married someone similarly placed it might well be possible to establish the young couple on a large farm. But most farms were small (only 35 per cent of the holdings were larger than 30 acres and 87 per cent of these were less than 100 acres), and there was an average of four or five children in each family. For most a suitable holding to occupy on marriage was a one-horse place or a one-pair place (such as Coedcoch and Llainmacyn respectively in the examples cited on pp. 106 and 112 above). In consequence in addition to those who were seeking suitable holdings when they intended marriage, there were young couples who had occupied one-horse places for some time and having accumulated capital were seeking larger holdings, there were families with growing children looking for larger farms that would give employment to the children as they came of working age, there were other families with many children of working age who were looking for still larger farms, and there were yet other families who were looking for smaller farms when their children had left home when they married.

This situation had direct consequences for people's interest in one another's family affairs, courtships, marriages, and links of kinship and affinity. At such a time of land hunger as the late nineteenth century and the early years of the twentieth century getting a farm was a difficult business. One South Cardiganshire landowner (Capt. Jones-Parry of Tyllwyd in Blaen-porth parish) testified in 1894 to the Royal Commissioners inquiring into the state of land in Wales and Monmouthshire, 'There has never been a death vacancy that has not been immediately applied for by several people; more than that, when people have been lying dead, before they have been buried, applications have been made for the farm, and I may say that before they have been dead I have had applications'.[10] Only by knowing about other people's family affairs could people who were themselves seeking farms know who else were courting and looking for farms, who were looking for larger farms, which farms were likely to fall vacant, and who had kinship connections which would help to secure farms for them.

For what transpires is that within groups of kin and affines individuals co-operated in order to accommodate one another in their movements from farm to farm. These examples are quoted from among the cases that have previously been cited:

1. A farmer (A) moved to a 110-acre holding when his father's brother (B) vacated it; he himself (A) was followed by his wife's father (C) in the holding that he (A) vacated. When A in turn vacated the 110-acre holding he was succeeded there by his wife's brother (D), who had succeeded to the holding previously held by his father (C), and which was then taken by a cousin of D's wife. When C's son (E) married, his wife's father vacated his holding and moved to a larger farm so that his daughter and son-in-law (E) could have a more suitable place in which to 'start their world'. While discussing such inter-connected movements (which span a period of many years) with an old man who had started work as a farm servant in 1883 he referred to the importance of being a member of such a group when he said, 'It wasn't impossible to get a farm, but it was difficult. It was like getting a job with the —— council, you had to have contacts.'

2. A moved into a 24-acre holding and vacated a 80-acre farm so that it could be occupied by B on his marriage, B being the son of A's wife's cousin. When A died C and D (who were B's sisters) moved into the holding that A had occupied until his death. E (the brother of B, C, and D) then became the sole successor to the tenure of the parental holding. When his son (F) married he (E, by

---

[10] *Royal Commission on Land in Wales and Monmouthshire, Minutes of Evidence*, vol. iii (London, 1895), p. 480.

then a widower) joined his sisters C and D at the 24-acre holding so that F could succeed to the occupation of the parental holding.

The following contrary example is provided. A and B were two brothers whose father held a 100-acre farm. A married the heiress of a 40-acre holding, no children were born to them and they spent their lives on that farm. B, the younger son, stayed at home with his parents and courted C for many years. They married when B's surviving parent (his mother) died. By then C was past the age of childbearing, and shortly after marrying B and C took a smaller farm that they could comfortably work without outside labour. They remained at the farm thereafter. Neither A nor B was faced with the problem of finding farms for children for they were child-less; they stayed on the farms that they had entered in A's case on marriage, in B's case soon afterwards. What primarily in-volved people in moving from one farm to another was catering for changing family circumstances; those whose circumstances did not change in this respect were less involved than other people were.

In the movements from farm to farm mentioned above and in others like them it is clear that precedence was not regularly given to any particular type of kinship relation and succession to farms was selective even in terms of kinship. That kin and affines were able to succeed one another in the occupation of farms owned by local gentry, as has been shown above, means that sitting tenants had a valuable perquisite, namely a voice in the choice of their successors, which they could exercise on behalf of their kinsmen. Not all kin were helped to get farms, and who it was who helped who was not a matter of kinship reckoning in a consanguineous sense but rather that some individuals were in a position to help some kin and affines, while others were not. This in turn was influenced by personal considerations and was dependent on such accidents of personal history as who was ready to marry at a particular time, who had to leave his farm involuntarily in the event of bankruptcy, or on the sale of an estate. It was a situation wherein mutual help to mutual advantage was part and parcel of the co-operation expected of kin whenever it was possible rather than that land holding should be regulated according to the genealogical connections of those in the occupation of farms at a particular time.

In conclusion it may be said that the way in which the society was ordered to meet the needs of working the land was not based on kinship, yet kinship involved individuals in one another's affairs in a way which made them distinctively one another's 'people'. While few men could meet the multifarious demands of daily life, social and psychological, year in and year out in independence of their fellows, it was the links of kinship which generally directed people's alignments and which guided people's preferences among their neighbours and acquaintances. From among the whole, kinship delineated for each individual a body of people who gave help at times of need and hardship and who sustained each other in the ordinary events of daily life and in the face of its demands. Those so connected constituted a pool of people to whom a man turned, and to whom his first loyalties belonged. The sentiments of kinship pervaded social life in one way or another; they 'informed' men's relations with all their fellows.

VIII

RELIGION

Introduction

BEFORE proceeding to see what light a discussion of the practice of
religion throws on the social structure of the rural community with
which this study is concerned it is necessary to draw attention to
certain features that have already been noted, for members and
adherents of religious bodies were people who were connected with
one another in the ways that have already been described. It will
be necessary too, to explain what the aim of this chapter is.

The community that has been under discussion was a 'face-to-
face' community, whose members were interdependent. This interdependence was necessary, and not merely a matter of convenience.
People were interdependent in farming operations, and again as
kinsmen, affines, and neighbours in times of hardship and exceptional circumstances. The means of transport available, at best
pony and trap, at worst on foot, limited the normal range of daily
journeys and limited the range of day-to-day personal contacts.
Again contributory to the nature of the community was the fact
that almost all people were members of farm work groups which
included farmers and cottagers, that farmers were members of
groups co-operating at the hay harvests, potato-lifting, and threshing, and that the opportunities for farmers' children to marry were
connected with knowledge about the availability of farms and about
other people's personal fortunes.

In practice the outcome was a knowledge of and interest in one
another's personal affairs, though this cannot be considered to be
simply a mechanical outcome of the circumstances that have been
noted. The situation may briefly be put by paraphrasing a defence
that Mr. Winston Churchill (as he then was) is said to have made
on one occasion. He was criticized for seeing that the House of
Commons was rebuilt after wartime bomb damage exactly as it previously was, and for failing to take the opportunity to modernize the
chamber. He is said to have replied, 'We shape our buildings and
our buildings shape us'. Similarly people shape their communities

and their communities shape them. Above have been noted some of the considerations to which people gave shape in their practices, and which in turn shaped them.

Thus those people who were members of, or who attended at, places of worship, were people who were already and on other grounds connected with one another. They were not simply a number of individuals without contact except in their worship, or in their attendance at places of worship. Overall they were stratified into gentry, farmers, and cottagers. The gentry, as has been noted, were members both of the local community and of a nation-wide stratum of society. As members of the local community they were landowners, responsible for the allocation of tenancies, for improvements in agricultural methods where they still had home farms, and responsible to an extent for the state of cultivation on their estates through the cultivation clauses contained in leases. As members of a nation-wide stratum of society they shared in that stratum's culture including its Anglicanism. In South Cardiganshire no member of the gentry was a nonconformist.

Farmers were connected with one another as members of co-operative groups, and by the prestige of being 'masters' *vis-à-vis* both their farm staffs and members of their work groups. They were connected with cottagers through the medium of farm work groups, in which the higher status belonged to the farmer and the lower to the cottager. All cottagers were in fact part-time farm labourers, but it has been seen too that members of different kin groups were not precisely divided into farmers and cottagers but overlapped that division.

The mass of farmers and cottagers were nonconformists, but some farmers and some cottagers were churchmen. In South Cardiganshire in 1901 there were ninety-one places of worship, one to each 186 head of population.[1] Chapel records show that people were usually received into full membership or full communion at the age of fourteen. Seventy-one per cent of the population was aged fifteen and over, so that there was one place of worship to each 132 people who had reached the age at which people were usually in full membership or communion.[2] In the parish of Troedyr-aur there were four places of worship, the parish church, one

---

[1] By 'South Cardiganshire' is meant that part of the county south and west of a line from New Quay to Llandysul.

[2] H.M.S.O., Census of England and Wales, 1901.

Congregational chapel, and two Calvinistic Methodist chapels. But people who lived within the parish regularly attended thirteen places of worship, nine of them being outside the parish boundary. (It should be noted that in this connection parish boundaries are of no significance to nonconformists.) In 1900 the average congregation of the parish church on a Sunday morning was thirty, and on a Sunday evening forty.[3] The Congregational chapel's membership was 233 (1909), one Calvinistic Methodist chapel had 187 members (1900), and an average Sunday school attendance of 104, while the second Calvinistic Methodist chapel had a membership of ninety-four (1907).[4]

At this point it is as well to recall that this is specifically a study of the social structure of a community, and this intention gives the study its direction and its character. The further discussion is not intended to be a study of religion *per se*, but of what is entailed by the data that have been presented above for the social structure which is the subject of this study. The nature of the study not only directs the aims but influences the terms of this inquiry. Men's conscious actions can be interpreted in terms of personal motivation, as when personal differences within one chapel lead to a group of members seceding to found another chapel. But the founding of a new chapel may also be interpreted in specifically religious terms, for example in terms of credo. When the Lloyds of Bronwydd established *Trinity* Chapel (Capel Drindod), a Methodist chapel, on their own land in the closing years of the eighteenth century it appears that one of their concerns was to stop the spread of Unitarianism in the Teifi valley. Clearly these possibilities are not alternative types of interpretation such that if one is true the other is not. Credal differences may be the basis of personal disagreements that lead to secession. It is not these considerations that are of primary concern here. Rather the intention is to see what understanding of this society can be gained by considering it in sociological terms. Lastly it is not claimed that such an understanding is in any sense more basic than other ways of understanding the material that is to be discussed.

---

[3] Church in Wales Records, SD/QA/54–5, in N.L.W.
[4] It is not possible to compare directly figures for a nonconformist chapel and those for a parish church. This is because that when people were admitted to full membership of a nonconformist chapel they became members of that particular chapel, while those received into the communion of the Church were received into the Church as such rather than into a particular parish church.

Relationships within a chapel

In considering relationships between the members of a chapel consideration will first be given to the constitution of the membership in the light of what has been said above. It will then be possible to consider the status system that existed as between the members.

Membership of a chapel, I speak metaphorically for purposes of convenience, divided those who on grounds that have already been considered (ecology and kinship), were associated. It divided farmers from farmers and it divided cottagers from cottagers. In no case was one chapel in a locality frequented by farmers and another chapel by cottagers. Thus in the village of Rhydlewis in Troedyraur parish, which has two chapels (Congregational and Calvinistic Methodist) the farmers were divided in their membership, some being members of the one chapel and others of the other chapel. Cottagers were similarly divided, some being members of the Congregational chapel and others of the Calvinistic Methodist chapel.

To a degree nuclear families, sibling groups, and other 'near kin' and affines were similarly divided. The way in which this happened can easily be presented. When two people who belonged to different chapels or denominations were married, if they had previously lived in different localities then the one who moved on marriage to join the other also removed his or her membership to the chapel or church where the spouse was a member. This made husband and wife members at the same place of worship, but separated the one or the other from his or her parents and siblings. When two married who were of different denominations, or of different chapels of the same denomination, but who lived in the same locality so that neither had to remove from the locality on marriage, then on the first Sunday after the marriage they both attended the place of worship where one spouse was a member. On the following Sunday they both attended the place of worship where the other spouse was a member. Afterwards each attended at the different places where each was a member. On the birth of children some attended with the father, others with the mother. Thus nuclear families, sibling groups, and groups of 'near kin' were liable to be divided in respect of their allegiance to different denominations and different chapels of the same denomination. Two examples follow.

1. R.J. and J.D. were two brothers, the sons of Calvinistic Methodist parents, and themselves Calvinistic Methodists. R.J. married a Congregationalist who removed from her home ten miles

away to join her husband when they were married. She then joined the Calvinistic Methodist chapel where her husband was a member. All their children became members of that chapel. J.D. also married a Congregationalist, but she already resided in the same locality as her husband. She remained a member of the Congregational chapel. The sons of the marriage became Calvinistic Methodists with the father, the daughter became a Congregationalist with the mother. In the case of R.J. and his wife, the nuclear family was kept complete in respect of religious membership, but at the expense of separating (in this respect) one parent from her siblings. In the case of J.D. and his wife, the sibling groups of both parents were kept complete in respect of religious membership, but at the expense of dividing the nuclear family both in respect of the parents and of the siblings.

2. The second example concerns two sibling groups, all their members being Calvinistic Methodists initially. T., who was the eldest brother in one of the sibling groups, was married to M., who lived in the same locality. Both remained Calvinistic Methodists. They had two children, both Calvinistic Methodists like their parents. One of the children removed from the locality on marrying a churchman, and was herself received into the communion of the Church. D., who was T.'s brother, married a churchwoman who lived in the locality; the one remained a Calvinistic Methodist and the other a churchwoman. They had two children, one was received into the communion of the church, the other became a member of the Calvinistic Methodist chapel. L. was the third brother in this sibling group, and he was married to a churchwoman who lived in the locality: two children were born to them, and both became communicants of the Church after their mother had been widowed while the children were still young.

E. was the sister of M. (who had married the eldest of the brothers in the sibling group mentioned above). She married a Congregationalist who lived some six miles away, removed there to join him, and became a Congregationalist. She was widowed when the two children of the marriage were in their teens. She thereon returned to her native locality and rejoined the Calvinistic Methodist chapel where her sister (M.) was a member. Her elder child, having been received into full membership of the Congregational chapel in the locality where he had been born and reared, remained a Congregationalist. Her younger child, who had not been so received, attended the Calvinistic Methodist chapel with his mother, and was later received into full membership.

These examples illustrate several points. Changes of membership of places of worship on marriage, and the division of nuclear families, sibling groups, and other groups of 'near kin' in respect of their religious allegiance involved all denominations including the Church. These groups were on occasion divided too in that their members were of the same denomination but of different chapels within that denomination, and this included cases where the husband was a member of one chapel and the wife a member of another near-by chapel of the same denomination. Where husband and wife came from the same locality each on marriage remained with the chapel (or church) 'where he (or she) was reared', so that attendance at one's chapel renewed the personal contacts made in the past in that way. Where two people were members of the same chapel or church before marriage, no such divisions as have been mentioned above were involved on their marriage. Nor is it possible to 'read back' from the situation that has been described above to a period when families and kin groups were uniformly of one place of worship or of one denomination or another, for change of membership on marriage can unify the religious allegiances of members of families and kin groups as well as divide them. In 1901 there were 203 occupied houses in the parish of Troedyraur.[5] I have details of the religious adherence of 169 of these households (82 per cent), the nuclear family was divided in respect of its religious allegiance in one case out of every eight (20 out of 169, $12\frac{1}{2}$ per cent), sibling groups were divided in at least another $12\frac{1}{2}$ per cent of the households, and it was rare to find a group of 'near kin' without some division in the religious affiliations of its members.

Thus the constitution of the membership of chapels and churches was of what can be labelled 'sociological fractions' (in so far as the social structure has yet been considered). This can be further illustrated by referring to the members of the congregation of the parish church of Troedyraur in the opening years of this century. In 1900 the average congregation on a Sunday morning numbered thirty as has been said earlier, and on a Sunday evening forty. I have been able to trace thirty-seven people who attended the church regularly, and they were members of seventeen households. Nine of the regular

5 H.M.S.O., Census of England and Wales, 1901.

attenders were gentry and their servants. The tenth was a member of a household whose two other members were Baptists. The eleventh and twelfth were husband and wife; the husband had previously been a Calvinistic Methodist but turned to the Church after quarrelling with his siblings who remained Calvinistic Methodists. The thirteenth, fourteenth, and fifteenth were parents and child, a church-going family of farmers. The sixteenth and seventeenth were husband and wife, and again farmers. The eighteenth and nineteenth were also husband and wife, and cottagers. The twentieth and twenty-first were mother and daughter, cottagers, who attended both the Calvinistic Methodist chapel where they were members and the parish church on each Sunday. The twenty-second was a farmer's wife, while her husband and son were Congregationalists. The twenty-third was a spinster cottager living on her own. The twenty-fourth was one of five siblings, all unmarried and living as one farm household: the other four were Calvinistic Methodists. The twenty-fifth and twenty-sixth were a farmer and his wife: the husband had been a communicant of the Church but on removing to the parish of Troedyraur on his marriage to a Calvinistic Methodist he himself became a Calvinistic Methodist. He then quarrelled with a neighbouring farmer who was prominent in the Calvinistic Methodist chapel that he attended, thereon he returned to the Church and his wife accompanied him. The twenty-seventh was a cottager who shared the cottage with his sister, who was a Calvinistic Methodist. The twenty-eighth lived at home with his father who was a Congregationalist and his mother who was a Calvinistic Methodist, while the twenty-ninth was a professional man who was married to a Congregationalist. The thirtieth was the wife of a Calvinistic Methodist deacon, and the thirty-first was a cottager who lived on his own. The thirty-second was a farmer's wife, her husband being a Congregationalist. The other five were the wife and four children of a farmer who himself was a Congregationalist.

Several points need noting. As has previously been remarked all the gentry were communicants of the Church of England (as it then was). It has sometimes been said that it was the richest and the poorest that attended the church. Actually in this case the communicants of the Church included farmers, shopkeepers, and cottagers as did the nonconformist churches. It is to be noted too that in respect of the details given above the Church was not different from other denominations. Every case in which the religious allegiances of members of a household were divided involves a division in allegiance to the nonconformist chapels as well as to the Church, and personal differences affected the composition of the membership of all religious bodies, not merely that of the Church. Thus those who were associated with one another as farmers, as cottagers, as members of farm co-operative groups, as members of individual farms' work groups, and as groups of 'near kin' and affines commonly found themselves divided by their religious allegiances. What is of immediate concern are the relationships within religious bodies constituted as has been discussed above, of people living in a face-to-face community.

As membership of religious bodies was constituted in this way one consequence was that the main divisions and allegiances of the community were paralleled within each chapel, except that the gentry were not communicants of any denomination but the Church of England. This apart the main features of relations outside the chapels were reproduced in them. The division between farmers and cottagers was clear and sometimes bitter. When the members of one South Cardiganshire chapel were discussing the possibility of building a manse for the minister one cottager spoke at great length so that others had little opportunity to express their opinions. When at last the cottager finished speaking a farmer spoke, saying 'Now that you "wheelbarrow men" (every cottager had a wheelbarrow for use in his garden) have had your say, perhaps we "cart men" may have ours': the distinction thus pointed was obvious enough to those present. That this division was sometimes bitter can be seen in the words of a Teifi valley minister of forty years' experience who was known for his level-headedness and who was President of the Welsh Congregational Union in 1904: 'There are different classes in the church, men vary in their circumstances, in their positions, in their talents, and in their sensibility. A man's circumstances give him standing, and his standing gives him influence. The poor are jealous of the rich, and it is no easy matter to be fair to both. . . . There is a danger of founding our respect towards one another on worldly standing.'[6] In 1905 he wrote in the denominational journal, 'Satan takes advantage of the great lack of sympathy that is to be seen in some places between the different classes within the churches. We know of churches wherein differences in circumstances and positions in life have endangered

[6] B. Davies, *Y Pulvud a'r Seddau* (Dolgellau, 1909), p. 125.

the peace of the churches, and almost destroyed the co-operation of their different members.'[7]

The division in the society between farmers and cottagers was not new, it was present at least sixty years earlier.[8] So was the land hunger which led to difficult relationships among the farmers themselves, but the situation was aggravated at the end of the century for then estates were being sold. On occasion farmers purchased farms over the heads of the sitting tenants (examples have been given in Chapter V above) which led to bad feeling between the parties concerned for this practice was generally condemned. But the most serious quarrels, which led to lasting animosity, were not quarrels about the occupation of farms or quarrels between members of different kin groups but quarrels between members of the same family, more particularly between siblings. It cannot be said that there were any such institutionalized customary procedures for reconciling such quarrels, as are to be found in some societies. Rather the quarrels were expressed by those who had quarrelled ceasing to be on speaking terms, withdrawing from one another, breaking contact, with one party removing his membership from the place of worship frequented by the other party to the quarrel. But it might express itself too by the two parties taking opposite sides over public issues, and places of worship were among the places where public issues were discussed. Thus on the occasion when a chapel was considering 'calling' a new minister, members who were friendly with one party to a quarrel, on speaking well of one minister who was being considered, were accused by the other party to the quarrel (the brother of the first party) of bringing pressure to bear on others to support them, and their personal integrity impugned. The chapel might become the public arena of private quarrels. Not infrequently I heard it said regretfully that 'one must take the quarrel to the chapel' (*rhaid mynd a'r gynnen i'r capel*).

Certain chapels were acknowledged to be difficult ones to minister to because of the personal rivalries between members, and many of these rivalries arose from 'structural situations', situations that were a necessary consequence of the social structure. Thus in the above example the occasion of the quarrel between the two

---

[7] *Y Tyst* (Merthyr Tudful), 6 Dec. 1905.
[8] D. Williams, *The Rebecca Riots* (Cardiff, 1955), pp. 243, 268. See H.M.S.O., *Adroddiadau Dirprwyaduron Ymholiad i Gyflwr Addygiad yn Nghymru*, vol. ii (London, 1848), pp. 151–2, for a statement made by the curate of Troedyraur parish relating to the conditions in that parish.

brothers had been that one (a married farmer) considered that his unmarried brother (who was also a farmer) left at home with their widowed mother was not fulfilling his customary obligations to her. While what has been said here is valid it is as well to record that personal friendships and attachments were expressed in places of worship as well as personal differences, and that they were places of friendly rivalries for chapel offices and for expressing one's opinions as well as for less friendly rivalries.

In such situations the position of the minister himself was that of a man who was, or ought to be, above party, with the role of an 'institutionalized outsider', free of the connections that would render his standpoint suspect. It was virtually unknown for a chapel to 'call' a minister who was a native of the locality, and he therefore came there uncommitted to any group of kin or any particular party. There was concern in any chapel too if the minister should marry a member (and stay in the locality) as he was then considered committed to his affines. Nor should a minister connect himself with some families or households rather than others among the members of his chapel. One very prominent and successful minister was described to me thus: 'He was a real old fox, he did not stay long in any house', that is, he showed no favouritism. This was not said slightingly but to express the fact that the minister understood perfectly well what the situation was and conducted himself accordingly.

Very rarely the division could and sometimes did become irreconcilable, so that one party might leave to become a separate congregation and build its own chapel. In one case where this happened the personal differences between members were such that though the overt issue was simply where the minister should reside (he was actually residing nearer to another chapel than to the one to which he ministered), the parties came to actual fist-fighting before one party seceded to establish a separate congregation and a new chapel. But such happenings were quite exceptional. Of the sixty-three nonconformist chapels in South Cardiganshire only four were the result of such 'splits'. Normally the parties had to accommodate themselves to one another, and either one or both accepted the discipline imposed by the body of the congregation and the minister.

The status system within the chapel

In *The Social Sources of Denominationalism* H. R. Niebuhr has considered one extreme case of what is liable to happen to the religious organization of a country in which there are marked social divisions, whether on class lines as in Great Britain, or between immigrant groups as in the United States until the early years of the twentieth century.[9] This, granted religious toleration, is that different social strata, or different groups, all come to possess their own churches, as those who consider themselves in an inferior position (in the community or in the existing church) secede to establish new churches. Niebuhr labels these 'the Churches of the Disinherited'.

In that the gentry (who were numerically a very small proportion of the population) were communicants of the then Established Church, the Church of England, and none was a member of a non-conformist denomination, the situation in South Cardiganshire (and elsewhere in Wales) may be interpreted in this way to a degree. Indeed one aspect of the growth and development of nonconformity during the nineteenth century was that the common people were dissociating themselves from the gentry, and that it was in and through the nonconformist churches that a new leadership was rising as is discussed below. But only to a limited degree can the situation be so interpreted. Whereas Niebuhr describes successive secedings as a 'Church of the Disinherited' is itself cleaved by the secession of the more disinherited within that church, there was no comparable development in South Cardiganshire. Farmers, cottagers, and professional people remained as members of individual chapels. There was nothing comparable to what Jeffries noted in some parts of southern England where cottagers were virtually the sole members of nonconformist churches while virtually all the farmers remained in the Established Church.[10]

The other extreme case has on occasion been reported by anthropologists from among peasant societies, wherein relations between individual members within the church (or corresponding religious organization) are the same as the relations between those same individuals without the church, so that those who are of high standing in the social fields based on ecology and kinship are also of high standing within the religious organization and conversely those who are of low standing in the social fields based on ecology and kinship are also of low standing in the religious organization.[11] As

[9] H. R. Niebuhr, *The Social Sources of Denominationalism* (New York, 1957).

[10] R. Jeffries, *The Toilers of the Field* (London, 1898), pp. 104–5.

[11] W. J. Goode discusses variations in the relationship of status in the secular field to that in the religious field in his *Religion among the Primitives* (Glencoe, 1964), in which the discussion is not limited to primitives.

status relations within the church correspond to those outside the church it has then been possible in such cases to consider the whole as a unitary social field, and to regard the religious organization as underwriting or validating the values of the social fields based on kinship and ecology.

A tendency to accommodate relationships within the church or chapel to those of the social fields based on kinship and ecology was clearly to be seen in South Cardiganshire. Far more farmers were elected deacons than their numerical strength would suggest. It was well known that certain of them had little to contribute in regards of 'public gifts', that is ability to speak or pray at religious meetings, or to contribute to church administration. Some had achieved their positions 'through their pockets' (that is by substantial contributions to a chapel's funds), and in parts of North Pembrokeshire it was said that a farmer should own a pair of horses before he was qualified for the diaconate. Craftsmen commonly occupied the less elevated positions in the chapels as teachers in Sunday schools and singing classes. At the teas that accompanied many functions in chapels and churches it was the cottagers' wives and daughters who tended the fire, boiled the water, and washed up the dishes while farmers' wives and daughters tended table, with a good deal of rivalry as to who should tend the ministers' table at meetings which brought a number of ministers together. Fifty years and more after the occasion one lady described the embarrassment she had felt on realizing that she had once inadvertently asked the daughter of a substantial farmer to help with the washing up.

But this tendency to accommodate relationships in the one field to those in the others was never complete, for there was another tendency which is discussed later. That it was not the case that status relationships in the religious organizations had been completely accommodated to those in the field of ecology is clearly seen from these details of the diaconates of the three nonconformist chapels in the parish of Troedyraur at the turn of the century:

Chapel I   Deacons
              Shoemaker
              Schoolmaster
              Farmer
Chapel II   Miller
              Farmer
              Farmer

Chapel III Farmer
        Farmer-merchant
        Shoemaker
        Carpenter
        Labourer occupying a one-horse place. (On
        one occasion a public collection was
        organized to aid him when he was in
        straitened circumstances.)

Meanwhile in each of these chapels there were many farmers who were not deacons. Consequently there arose a possibility of an incompatibility between a person's status and duties in the one field of relations as contrasted with his status and duties in the other field of relations. A nineteenth-century example makes this point clear. The farmer and chief servant of a 253-acre farm were members of the same Calvinistic Methodist chapel; the chief servant was a deacon while his master was not. It fell to the chief servant to reprimand his master for misdemeanour, in front of the congregation at a meeting of the chapel's religious society, and the master accepted his disciplining at his servant's hands. There was a radical incompatibility between the status occupied and the role played by each in the social field based on ecology and the status occupied and the role played by each in the religious field. This implies the common recognition of another authority than that which obtained in the field of relations based on ecology, and however partial its acceptance upon occasion it was not a temporary phenomenon. It persisted in nonconformity for over two hundred years. The problem of the relationship of those who were very unequal in 'the world' but equals or with their relative standing inverted within the chapels faced dissenting churches of the seventeenth century.[12] It faced the newly established Methodist societies during the eighteenth century, and was explicitly discussed by one of the most important of the Welsh Methodist leaders William Williams of Pantycelyn.[13] It was still a problem in the late nineteenth and early twentieth centuries. Two centuries and more did not fully accommodate the one field of relationships to the other.

[12] This can be seen in the details contained in the covenant of the church of Mynydd-bach, dated 1700, and published in T. Rees and J. Thomas, *Hanes Eglwysi Annibynol Cymru*, vol. ii (Liverpool, 1872), pp. 4–8.
[13] Williams discusses this in 'Drws y Society Profiad', *Holl Weithiau Prydyddawl a Rhyddieithol y diweddar Barch William Williams* (ed. J. R. K. Jones), (London, 1867), pp. 651–70.

It was not for lack of time that the accommodationist tendency was only partially effective, there was a more basic reason. It was implicit in the situation because of the incompatibility of the values of the two fields.

When the religious revival of 1904 and 1905 came it affected the inhabitants of both the agricultural and the industrial areas of Wales, and as the revival is discussed below a comparable example to the above from the industrial area of East Carmarthen and West Glamorgan is here provided. Prior to the amalgamations of the 1920s the works in this area were commonly small in size, and the social structure of the industrial villages in some ways not unlike that of agricultural communities in south-west Wales. The owner (or part owner) ran the works, and was known to the workmen just as a farmer was known to his farm staff and work group. The 'boss' lived in the same area as his employees, and not only because of the limitations that the transport facilities available placed on people who had to attend the works daily. The contacts between the employer and the employees were not limited to the works, and in the case that is discussed here the employer was a deacon in the chapel where a number of the workers that he employed in a small steel works were members. There was a strict rule in the works that any stoker who let the furnace die out while the plant was idle at night or over the week-end faced immediate dismissal. One stoker, a married man with a family, did allow the furnace to die out, and was dismissed. Thereupon a deputation of his workmates who were members of the chapel where the employer was a deacon, approached the employer to try and get the dismissal revoked as the stoker's family faced hardship. When the 'boss' proved adamant, the deputation's spokesman announced that only one thing was left to them. Thereupon the workmen knelt and prayed in the employer's presence. The stoker kept his job.

As in the earlier example there was a common acceptance by those of unequal rank in 'work relations' of values and ideals and obligations which a man could not deny in practice without revealing himself as the sort of person that the community derided. On occasion that did happen, and then the person who had demeaned himself was commented upon ironically 'What a Christian!' But the contrary happened too, as in the above examples. Common acceptance of values and ideals for which a validity independent of 'the world' was claimed and accepted, surmounted the differences and distinctions which existed in the field of relations based on ecology or the 'works relations' of an industrialized area, as the case might be.

It has been noted above that in some communities the status a man occupies in the religious field corresponds to that which he occupies outside the religious field, and that this correspondence allows one to think of the whole as a unitary social field. While a tendency to accommodate a man's status in the one field to his status in the other can be seen in South Cardiganshire, it is nevertheless clear that this tendency was only one factor in the situation and that it did not come to be of exclusive importance. What is to be found here is not one unitary social field but two fields, necessarily related because both concerned the same individuals, but with independent bases. There was no necessary connection between statuses and roles in the one field with those in the other: there was always the possibility and on occasion the actuality of a person occupying statuses and performing roles that were incompatible with one another. People felt such a situation to be incongruous and went part of the way to resolve the incongruity by being readier to elect to the highest chapel offices those who were of higher standing in everyday life, farmers, merchants, professional men, people who, so it was claimed, had the experience of management that would help them to manage the affairs of a chapel. On the other hand a cottager lacked this experience and he would need to be outstanding in his 'public gifts' and character to gain election.

There were such cottagers and they were on occasion elected to the highest offices in the chapels but they were generally cottagers who were self-employed as craftsmen, or labourers whose working days were over. The values and beliefs and ideals taught in the chapels claimed a validity and an authority independent of 'the world'. It cannot be claimed that this teaching 'validated' or underwrote the way in which this society was organized to meet the needs of working the land. Rather it emphasized ideals and obligations that were incumbent upon all regardless of their worldly standing, the virtues of family life, the obligations of industriousness, the duty of developing a certain type of moral character, and of faithfulness in striving to reach an ideal that was part of the cultural heritage of the several denominations. Chapels too were dependent on the cultivation of the 'public gifts' of their members, as speakers, teachers, and at prayer. Where these qualities were recognized in a man the authority of the values and ideals involved were recognized by election to high office regardless of his status in 'the world'. But this was only possible, granted that the ideals of the religious teachings were based not on the detailed configuration of this particular ecological field, but were held to be valid independently of it.

Historically too the origins of the values and beliefs of the religious bodies were quite independent of the way in which the society was organized to work the land. One of the most prominent features of Dissent has been that it has incorporated a particular notion about the nature of a church. The early Dissenting churches were 'gathered churches' whose members covenanted 'with God and with one another' to respect the detailed arrangements contained in each particular covenant, and such a notion about the nature of a church was inseparable from other theological ideas, such as the nature of the priesthood. These notions did not arise in South Cardiganshire or in Wales or in Britain, but on the mainland of Europe and reached Wales by way of England, and individuals were persuaded of their validity. They were not founded on the existing society of South Cardiganshire and they retained an authority which was independent of the way in which that society was organized to work the land.

Relationships between chapels

It has been seen that in South Cardiganshire there were in 1901 ninety-one places of worship, one to each 132 people who had reached the age at which people were usually received into full communion. It has also been seen that the inhabitants of the parish of Troedyraur attended regularly at thirteen places of worship, four within the parish and nine without. It has been noted that membership of a chapel or being communicants of the Established Church divided those who were on other grounds associated with one another, so that membership consisted of 'sociological fractions'. Thus the harmonies and divisions and disunities which existed outside the places of worship were brought into them. But having divided those who were on other grounds associated, membership of churches and chapels then produced a measure of unity between the 'sociological fractions' of which each was constituted, through the common acceptance of values and obligations, and through another and different medium which while it united the members of individual chapels, at the same time divided one chapel from another, namely competition, not within but between the different bodies of worshippers.

In general this took such forms as which chapel was the best kept, which was most imposingly built, which had the best grounds, which had the largest membership, which the largest congregation. People took pride in their chapel and its prestige, and in the attainments of its members. If one chapel secured the services of a noted preacher for its annual preaching festival, the other chapels wanted to secure equally noted or better-known preachers for their preaching festivals. Those who might be disunited within a chapel were drawn together *vis-à-vis* other chapels. If a chapel was to be known for its flourishing congregation, the success of its services, its achievements in its social and cultural work, its members had to sink their differences and co-operate. And its success was judged not only by the achievements of its individual members in organized periodic examinations in biblical knowledge and on other such occasions, but as contrasted with the accomplishments of other chapels on set occasions which brought the members of various chapels together, as is noted later.

Here are provided two examples of this competition which in these cases was seen on the occasions on which the 'school marched'; the 'school' was the Sunday school (of a particular chapel) and its members marched in public on the day of one of the Sunday school's festivals. At New Quay the schools of both the Calvinistic Methodists and the Congregationalists marched, separately and on different dates. Each marched on the day of its annual Sunday school tea party. Both marched at the same season of the year when the hay harvest was over. The tea party was held in a convenient field, but the members assembled not at the field but at the chapel where the school's officers arranged the order of the march. Leading the procession were two people who carried a banner that bore the chapel's name, and alongside them the minister walked. The deacons followed the standard-bearers and they were followed by the youngest children with the girls in front and the boys behind. Their Sunday school teachers walked alongside them. Behind these came the 'young people', with the adults bringing up the rear. Everyone carried a mug hanging around his neck. The procession having formed up outside the chapel, the precentor led the singing of a hymn before they moved off. At certain prearranged points along the route the procession halted and a hymn was sung while members of other churches watched the procession and listened to the singing. On reaching the field, which was lent for the occasion by a local farmer, the tea party followed and when it was over the procession reformed and marched back to the centre of the village before dispersing. For days afterwards the procession was discussed, and especially the singing, while arguments proceeded as to whether the singing was better at any one station than at another. In a matter of days the other chapel's Sunday school held its procession and followed the same routine. Then followed prolonged discussion as to which procession had been superior, and as to which school had sung the better, whether the one had excelled at some halts and the other at other halts. Each school was particularly concerned to shine and to maintain the good name of the chapel.

For schools to march in this way was characteristic of industrial areas rather than agricultural areas, but a second example follows from North Cardiganshire. Taliesin and Tre'r-ddôl are two small villages, standing less than a mile apart. The former has a Calvinistic Methodist chapel and the latter a Wesleyan chapel. After the midday meal on Good Friday the members of each congregated separately at either chapel, and at each a procession formed up. In front came two young men carrying the chapel flag. These were followed by the children each carrying a small flag with a painted handle. Then came the 'young people' and lastly the adults. They then marched through the village in which they had assembled, to the far end of the other village, and back to the vestries of the chapels whence each had proceeded. While marching they stopped outside the house of each member who was ill and unable to attend, and there sang a hymn. On returning to their starting-point, a tea party awaited each procession in either vestry, to be followed in each case by a concert. At the tea children were served first, followed by the men and then the women. When the tea was over items of food were set aside for presentation to those unable to attend because of illness and the room was then prepared for the concert. This consisted of items of singing and recitation first by the children and then by the adults. For days afterwards people debated which chapel had proved superior, which had the best procession and hymn singing, and which had the better concert. (This was judged partly by the length of the concert, in which respect one chapel had the advantage of a gramophone which was played during the evening.) As in the first example members took pride in contributing to their chapel's success, on this as on other occasions. It was noted above that such processions were characteristic of industrial areas rather than agricultural areas, and one who took part in a procession in Glamorgan later recounted, 'My greatest contribution to that *Cymanfa* (festival) was to act as one of the flagbearers of the Nebo standard, and very important service did I, at the time, consider it to be'.[14]

In South Cardiganshire the occasions of the major competition between chapels were events which brought the members of several chapels together at the same time and place. Each chapel possesses several societies which are open to its members, but formal membership of the chapel does not necessarily make one a formal member of these internal societies. In addition to being a member of a chapel one may also be a member of its Sunday school, and one may further be a member of its temperance society. In practice membership of these internal societies tended to be coterminous with the membership of the chapel.

In consequence, as events were arranged which brought together the corresponding specialized societies of different chapels on separate occasions, in practice various chapels met together not once but several times during the year. The two most important of these occasions were the singing festival and the 'subject' festival (when the Sunday schools were examined on passages of Scripture that had previously been chosen as the *subject* of study). The Calvinistic Methodist Church (now known as the Presbyterian Church of Wales) has a formal presbyterian organization wherein chapels are grouped into presbyteries both for church government and for functions such as singing festivals and subject festivals. As a result each of the two Calvinistic Methodist chapels in the parish of Troedyraur convened with some five other chapels to hold each of these two festivals. The Congregational Church does not have a presbyterian organization but it is long established that the Congregational chapel in Troedyraur parish meets with five other Congregational chapels to hold their singing festival. The Congregational chapels are grouped slightly differently for the subject festival, and to this end the Congregational chapel in the parish of Troedyraur meets along with four other Congregational chapels. Certain Church of England Sunday schools also maintained a subject festival, thirteen of them

14 J. Ballinger, *Gleanings from a Printer's File* (Aberystwyth, 1928), p. 59.

meeting at Llandysul each January to recite their chosen scriptural passages, to be examined on the content of the passages, and to sing anthems. Another eight Church of England Sunday schools met at Newcastle Emlyn to the same end in late April.

The date of the festival, the venue, and the scriptural passages to be studied having been decided aforehand, the subject festival when it came was a day-long affair. While the Congregationalists customarily met at the same chapel every year, the Calvinistic Methodists met at each of the presbytery's chapels in turn, and this is now described. Each host chapel was anxious that the festival should be a credit to it. People compared it with other festivals held at other chapels in other years, with respect to the state of the chapel and its grounds, the food prepared for visiting schools, as well as the catechizing which was the main overt subject of the festival. During the morning the children recited the relevant scriptural passages from memory and were then catechized upon them. The adult schools' turn came during the afternoon. The whole congregation sat on the ground floor of the chapel, with the school from each chapel in turn ascending to the gallery to be catechized. Having mounted to and occupied the gallery in solitary state, and in full view of the other schools present each school first recited the scriptural passage that it had chosen in a manner that was characteristic of these festivals. Taking its cue from a leader the passage was recited in a rather staccato style dependent on a very careful observance of punctuation marks in the text, at a faster pace than in common speech, in a monotone and in a rather high pitch, so that the recital approached to a very distinctive stylized chant.

In some Baptist and Congregational festivals the recital was turned into something approaching a religious drama when the words of different Biblical characters in the text were recited by the different classes within the school. In one mid Cardiganshire village (Llanddewibrefi) in 1896, 'certain schools met to perform a *pwnc* (subject) on Job. The weather happened to be unusually bad, and a whole class of elderly women representing the messenger of woe failed to turn up, except one old woman. The effect was most realistic when the solitary representative of the class, in her old fashioned way, recited her part, "And I alone am escaped to tell thee"'.[15]

The word *pwnc* which we have translated 'subject', that is a

15 J. Ballinger, op. cit., p. 53.

subject for recital and discussion, also means a 'knotty question'. 'To be or not to be?—that is the question?' = 'Bod neu beidio bod —dyna'r *pwnc*', and after reciting the chosen passage of Scripture each school was questioned by the catechizer on the contents of the passage involved.[16] Some catechizers were particularly gifted in this respect and the success of the questioning, which was in fact a disputation rather than the answering of set questions, depended in considerable measure on the catechizer or questioner. The catechizing or questioning resolved itself not only into a competition as to which school recited and answered best, but into a competition between the members of each school and the catechizer. Catechizers did not confine themselves to questions of fact relating to the passage recited but concerned themselves with the theological issues involved and in this way the catechizing became the means whereby new theological ideas were placed before the chapels and their contents discussed. In this regard no minister was more important nor better known as a catechizer than the Revd. David Adams, B.A., who was awarded an honorary D.D. by the University of Wales but died before the honour was conferred upon him. Through the medium of catechizing individual schools in their own chapels (*ysgol ateb*) and a number of schools at the festivals here under discussion, he attacked the Calvinist positions while he presented what were then advanced liberal theological views to the schools both in his catechizing and in his extensive writings. The clash between able members of the schools who were of a Calvinist persuasion and such a catechizer resulted in furious debate which was very clearly regarded as a competition between the school and the catechizer.

On occasion he (the Revd. David Adams) questioned the school very searchingly, and the harder the conflict the better he liked it. He never seemed more pleased than when the whole school opposed him. In one festival a particular school received very harsh treatment at his hands, and all admitted that on the day he was the victor. But there is comfort in every event: that school's ablest debater was unable to attend that festival because of illness, which demanded that battle be rejoined on another occasion.[17]

In addition to reciting a scriptural passage and answering the questions asked about it each school sang an anthem of its own

---

[16] Ibid., p. 52.
[17] E. K. Evans and W. P. Huws, *Cofiant y Parch David Adams* (Liverpool, 1924), pp. 67–8.

choosing immediately after it had been questioned. If the questioning was regarded as a competition between schools, the anthem was considered an even greater competition, and one of the first questions asked by anyone who had been absent was 'Who took the singing?' (*Pwy aeth a'r canu?*), virtually 'Who won?'. There is no doubt that the people who took part regarded it as a competition. Many of them said, 'It was a real competition'. A celebrated local schoolmaster (John Newton Crowther) in later years recalled the remarks of an elderly deacon who had listened to one school singing an unusually elaborate anthem that included a number of solos and duets—'There they are! on "sharing terms" (*hannera*) today again! Pretending to sing to the glory of God, but all the praise to themselves.'[18]

The singing at the festival, like the reciting and answering, was the culmination of weeks of preparation in the singing practices that followed the Sunday evening services at each chapel. So much was the singing of the anthem a competition between the different chapels that in the eyes of many absence from chapel on a Sunday evening was reprehensible more because it involved missing the singing practice than the service. Each chapel had to cultivate its members' abilities in order to acquit itself well, and individuals who had much to contribute as teachers and singers were highly regarded whatever their position outside the chapel, for members took pride in their chapel's achievements.

This element of competition between chapels in respect of their ability to sing well at such festivals or conventions can be seen in this account of how a choir was established at the small market town of Newcastle Emlyn. Tommy Morgan was the precentor at Bethel, the Calvinistic Methodist chapel in that town, and when he had succeeded in greatly improving the standard of singing at Bethel, then

Having achieved so much, the next step was to build up a choir worthy of representing Newcastle-Emlyn, and to this end invitations were extended to singers not connected with Bethel to join in forming such a body, but there was for a long time no response. It seemed to be felt that as most of the singers were connected with Bethel chapel, it would always be regarded as a Bethel choir, and of course denomina-

---

[18] *Cardigan and Tivy-Side Advertiser,* 13 May 1921. The competition between chapels on these occasions is also noted in E. Evans, *Cofiant John Thomas, Llanwrtyd* (Caernarvon, 1926), p. 32. The subject of this biography was a native of South Cardiganshire. Competition between chapels in another area is noted in W. Phillips, *Mor las oedd fy Llannerch* (Caernarvon, 1955), p. 95.

tional spirit had its effect. It was probably to overcome this feeling that Tommy Morgan began rehearsing some of the masterpieces of the great German composers in English, so that it might not be felt that the Bethel choir would afterwards use them in their own Sunday School Conventions.[19]

On the other hand in a North Pembrokeshire district within the Teifi valley where the Baptist churches did not hold a subject festival the best vocalists of a Baptist chapel helped their Congregational neighbours to shine at their subject festival by joining them to sing the anthem. This was an occasion which did not bring the two chapels into competition with one another, while on occasions on which they were in competition there was no such co-operation.

While the subject festival was the most important occasion of the year for the disputation and competition between chapels which has been described above there were other occasions when smaller numbers of chapels met to the same end, by agreement. Thus the Calvinistic Methodist and Congregational chapels in the village of Rhydlewis in the parish of Troedyraur met annually on Christmas Day, each school being questioned by the other's minister, and each school sang its own anthem. But later they combined to sing the anthem, and the man who brought this about occupied an interesting position in the community, that of a man in a measure independent of both causes and with something of his own to contribute. He was a Yorkshireman named John Newton Crowther (1847–1928) who at the age of nineteen and a monoglot Englishman was appointed headmaster of the British Society school in the village of Rhydlewis and later of the Board school established there. He learnt to speak Welsh fluently, and was both an able schoolmaster and a man with exceptional gifts to teach music to adults. His position as a schoolmaster made him independent of the way in which the community was organized to work the land, and this with a background that did not connect him with any one party in the community made of him what was above called an 'institutionalized outsider'. Refusing to teach the two 'schools' independently, he combined them into one choir.[20] Again when there arose the delicate matter of reallocating the pews in a local chapel, a business that Crowther called 'the devil of the seating places', it was he who was chosen as the secretary of the meeting called to undertake the work, and to implement the

[19] *Cardigan and Tivy-Side Advertiser*, 23 Apr. 1909.
[20] Ibid., 6 May 1921.

decisions taken there.[21] People who could and did stand outside the community, to a degree, had a function to perform in the community, as is discussed later.

The *pwnc* (subject) as it has been described above probably had various origins, among them the practice of the Revd. Thomas Charles (1755–1814) in catechizing Calvinistic Methodist Sunday schools that he visited in and around Bala on topics that he had

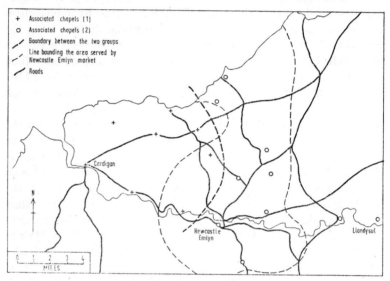

Fig. 13. Calvinistic Methodist chapels grouped together for various ends, 1900.

outlined to them in advance and had bade them prepare. But it appears that it was in south-west Wales that this form of religious instruction had the strongest hold, where it has lasted until today.[22]

The other major occasion that brought several chapels together was the singing festival. As has been noted the Congregational chapel in Rhydlewis joined with five other chapels to hold this festival: both of the two Calvinistic Methodist chapels in the parish of Troedyraur each joined with some five others to hold their festival. While the group of Congregational chapels met at the same centre every year, the Calvinistic Methodists met at different chapels annually (except at those which were too small to house the con-

[21] *Cardigan and Tivy-Side Advertiser,* 9 July 1920
[22] J. Ballinger, op. cit., p. 53.

gregation). Again there was a great deal of discussion comparing the festival with the past festivals, and with festivals held in other chapels. The children's festival was held during the morning and the adults occupied the rest of the day. On this occasion the various chapels did not sing in turn but as one congregation and the element of competition was less emphasized than at the subject festival. But as shall be seen later, in another way it was equally characteristic.

Here attention is drawn to certain other aspects of these festivals and conventions that drew the members of various chapels together. The singing festivals were the culmination of a winter's work in singing schools, and during this period various chapels held joint rehearsals so that at all these functions people from a considerable area met and came to know one another. In addition to the subject and singing festivals, a third festival was held in early May, namely the temperance festival. This was undenominational and was held by a number of chapels grouped to this end. The river Ceri flows through the parish of Troedyraur, and the Ceri Side Temperance Festival was held by a group of eight chapels, meeting at different chapels in rotation. It consisted of an afternoon meeting in which the children recited a prepared text on 'Temperance' and were questioned on it, and an evening meeting when two speakers addressed the congregation on the subject of 'Temperance'. This festival was part of the work of the temperance societies of each chapel concerned, and throughout the winter temperance meetings were held at each chapel in turn, attended by delegates and members from each other chapel in the group. A representative from every chapel addressed each meeting, or contributed an item to the proceedings. Similarly each chapel had its own cultural society. In this connection chapels were again grouped together and meetings were held in rotation at each chapel in the group, attended by parties of 'young people' from each of the other chapels. Individuals were appointed beforehand to address the meetings each on his or her chapel's behalf. As a result of these continuing contacts and such major occasions as the subject and singing festivals people came to know one another and became integrated into an extended community. It is common today for people in their fifties who grew up when these activities were more flourishing than nowadays to know people who attended these functions within the geographical area covered by a group of chapels without even knowing the names of prominent people in chapels of a different denomination without the range of that group of chapels. Acquaintance with individuals thus achieved became the starting-point for further information, for with the acquaintance came knowledge about who he or she 'was', that is, who he or she was related to, what his or her antecedents were.

Before proceeding to another aspect of competition between chapels it is as well to make the point that many activities outside the chapels as well as within were seen as a competition. One feature of the area is the *eisteddfodau* (competitive meetings for poets, singers, and reciters) held throughout it. No one familiar with the work of many of the poets will underrate the quality of the best work done at the *eisteddfodau*, but for very many the main interest was less the quality of the work as the personal competition between the contestants. They commonly competed one year after another, at a circuit of local *eisteddfodau* and became regular opponents.[23] Such speculations were (and are) to be heard as 'A.B. beat C.D. last time. I wonder whether C.D. can manage it this time?'. Even those who had no intention of reading the verse written were interested in this competition. There was and there remains a similar interest in competition between choirs, and a readiness to regard soloists as the personal champions of localities or groups.

Nonconformity incorporated this feature into its arrangements. By the late nineteenth century that separation of nonconformists from the rest of society which had once produced solidarity among them had disappeared. 'There is not in these days the extraneous pressure which makes every church a compact and heroic phalanx.'[24] In South Cardiganshire once the opposition between the Church and the 'World' ceased to produce an internal unity within the churches, competition between chapels worked to the same end. It is noteworthy that when theological differences and debates divided the denominations during the first forty years of the nineteenth century their singing festivals were undenominational and joined them.[25] Once the theological differences between the denominations

---

[23] An instance of this is the circuit of *eisteddfodau* at which D. (Isfoel) Jones (1881–1968) Y Cilie, a celebrated figure at *eisteddfodau*, at one time competed regularly: Glynarthen, Rhydlewis, Blaenannerch (all held around New Year's Day), Llanarth (Easter time), Llandysul (midsummer), and Brongest. Source— personal information from Mr. D. (Isfoel) Jones.

[24] K. S. Inglis, *Churches and the Working Classes in Victorian England* (London, 1963), p. 71.

[25] R. D. Griffith, *Hanes Canu Cynulleidfaol Cymru* (Cardiff, 1948), chapter 3, and especially p. 71, where it is stated that the first festival to be held in South Wales was held at Blaenannerch in South Cardiganshire in 1808.

became less commanding the singing festivals as they developed after the 1860s became denominational and divided them. Whereas distinctive theological standpoints had previously provided foci of unity for the several denominations, competition between them now worked to that end.

While the above has been concerned with one aspect of the element of competition in the relationship between individual chapels, what follows concerns another aspect of this competition.

Such a festival as the subject festival or the singing festival was held by a group of chapels of one denomination. It was the occasion of competition between the individual chapels, but it was also the occasion of co-operation between them to provide a festival that would compare well with or excel the corresponding festival held by other chapels of a different denomination. These too were in individual competition, but co-operating with one another and considering themselves as one *vis-à-vis* the corresponding group of chapels of the other denomination. From one standpoint the chapels were separate units in competition with one another, from another standpoint they were all one unit in competition with a similarly constituted unit of another denomination. Different denominations held their festivals at the same season of the year, in late spring when the sowing season was over and the hay harvest not yet begun. And attendance at a particular festival was not confined to members of the denomination concerned. After each festival, and this was perhaps more particularly true of the singing festival, people commented on whether or not it was a good festival, whether as good as other festivals, and how it compared with the festivals held by other denominations. Clearly before this can happen there must be common ground between the 'competitors' (other than in a geographical sense) or comparisons would be meaningless, and this existed in the common cultural heritage of the different denominations, the fact that the Scripture was their common basis of discussion in subject festivals, and that they shared many of the same hymns, tunes, and anthems. It should be remembered, too, that there was common ground between them in that as individuals they were associated with one another in their day-by-day activities outside the chapels, as farmers, cottagers, relatives, and neighbours.

Nevertheless the competition and friendly rivalry were real enough and extended to the children. It has been mentioned that there is a Calvinistic Methodist and a Congregational chapel in the village of Rhydlewis in the parish of Troedyraur. The children who attended both chapels attended the same village school, and on the day before the one denomination held its festival the children who attended the other denomination's chapel spent their time on their way home from school killing caterpillars of a certain species to make it rain on the day of their rivals' festival, for there was a belief among children that killing this type of caterpillar would be followed by rain. If it rained on the following day people said jocularly, 'There you are, it's the chapel children who have done it'. Llwyncelyn and Ffos-y-ffin are two mid Cardiganshire villages standing less than a mile apart. There is a Congregational chapel in the former and a Calvinistic Methodist chapel in the latter. Children from both villages attend the same school at Llwyncelyn. Children who attended the one chapel killed snails on the way home from school the day before the other chapel's festival was to be held, again to bring rain and spoil the festival. There was a common notion among children that killing snails brought rain. At the week-night services the deacons reprimanded them for their conduct. Mention has been made of neighbouring chapels in North Pembrokeshire where the Baptists' vocalists strengthened the Congregationalists' choir when the latter took part in their denomination's subject festival. The Baptists did not hold a subject festival, but both the Baptists and the Congregationalists held singing festivals. Congregationalist children who lived along the roadway leading to the Baptist chapel stood outside their homes carrying opened umbrellas as the Baptists made their way to the chapel, to make rain and hinder the success of their rivals' festival. Baptist children acted similarly when the position was reversed. The children did no more than reflect the friendly rivalry that characterized their elders.

Given the conditions described above it was to the advantage of each chapel and denomination to cultivate the members' talents, and people appreciated the value of those who were able to teach and to lead in Sunday schools, singing classes, and cultural society meetings. It was also of advantage to people of any particular denomination to secure the appointment to such a position as local schoolmaster of a member of their own denomination, not because it would influence his school work but because of what he could contribute to the work and standing of his chapel and his denomination. But to say this is not to belittle the religious and cultural value of the work done under these circumstances nor the long and hard hours put in by devoted individuals in teaching children and adults. In the course of preparing for festivals children and adults learned passages of Scripture and hymns whose symbolism is rich and sustaining. That teachers were pleased when their charges acquitted themselves well in competition with others is not to say that the teachers were not primarily concerned with the personal benefit of those they taught and the intrinsic value of what they taught them. One of the ablest and most devoted teachers in the area was a lame craftsman who finished work early on winter week-nights at a financial sacrifice he could scarcely afford, in order to walk home laboriously and thence to the chapel to teach his class of children. Such people had a status in the society, a role to perform, and a distinctive standing. Their passing was no gain to the community.

*   *   *

The common way in which new chapels were established during the nineteenth century was by the separation by agreement of a body of members from an existing chapel. For those members who lived at a distance, a chapel commonly established Sunday schools that met either in a private house or in a schoolroom built for that purpose. Such a body had its own officers for its own limited purposes, but if the membership increased it was common to expand the work. Ministers were invited to preach there at special services while members continued to frequent the parent chapel's regular services. Ministers were invited too to catechize the Sunday school, week-night services were added, and members became aware of their corporate identity *vis-à-vis* the parent chapel. They came to feel that they were entitled to opportunities and benefits and offices conferred by an independent status. A Calvinistic Methodist chapel which in 1900 had a total membership of 187, and with an average Sunday school attendance of 112, had thirty-two Sunday school teachers, as well as two Sunday school superintendents, an office to every three or four regular attenders, and this in addition to the diaconate of three members. If members of a Sunday school that met in a schoolroom felt themselves able to support the financial burden, they sought to establish a separate chapel. The last chapel founded in South Cardiganshire was opened in 1894, but the same process was to be seen at work for at least the first forty years of the twentieth century.[26]

[26] There was no Roman Catholic church in South Cardiganshire in the period with which this study is principally concerned, and no account is taken here of Catholic churches recently established.

At the beginning of this century all the gentry of the parish of
Troedyraur (and neighbouring parishes) were communicants of the
Church of England: none was a nonconformist. Approximately
90 per cent of the common people who regularly attended places of
worship were nonconformists. In Troedyraur parish the Congregational chapel had 233 members (1909), one Calvinistic Methodist
chapel had 187 members (1900), another had 94 members (1907),
while the Church had an average congregation of 40 on a Sunday
evening (1900).[27] The gentry and the hierarchy of the Church of
England were very closely connected: Rhys Jones Lloyd was the
rector of the parish of Troedyraur while his nephew was Sir Marteine
Lloyd of Bronwydd who was the patron of six livings and in addition
had the most important voice in recommending appointments to
the living of Llangynllo (in which parish Bronwydd stands, adjoining Troedyraur). The Tylers had among them country gentlemen and
naval officers as at Gernos, and clergymen of the Established
Church. A. J. Bowen who was the owner of Troedyraur mansion in
1900 was a clergyman, and in his absence his son occupied the
mansion. The Bishop of St. David's from 1874 to 1897 was the son
of a Cardiganshire squire. The gentry not only worshipped at a
different church from the overwhelming majority of the common
people, and were closely connected with the hierarchy of the Church,
but they were socially distant from the common people who among
themselves were in close face-to-face relations. It was among the
common people that the competition between the nonconformist
chapels induced a measure of solidarity and this was sociologically
irrelevant to the gentry who were socially distant from them.

The point has been made that the membership of chapels consisted of 'sociological fractions' and that competition contributed
towards producing a measure of solidarity between them. It may
reasonably be asked whether one large chapel which would include
all would not have this consequence more effectively. It has often
been said that denominationalism with its multiplicity of chapels

---

[27] It is no longer possible to know exactly what percentage of the total adult
population were members of one nonconformist chapel or another, or communicants of the Established Church. But an estimate is possible granted one
assumption. Many of those who lived in the parish of Troedyraur were members
of chapels outside the parish and many who lived outside the parish were members
of chapels that stood within the parish boundaries. If it is assumed that these
balance one another roughly, then more than 90 per cent of all who were over
the age of fourteen were communicants of the Established Church or full members
of nonconformist churches.

has divided society in Wales, the assumption being that the divisions connected with denominationalism were simply additional to the other divisions that existed, those relevant here being the uneasy divisions into gentry, farmers, and cottagers.

What is suggested here is that in South Cardiganshire during the late nineteenth and the early twentieth centuries when conditions of agricultural depression and the sale of estates contributed to difficulties in the relationship between farmer and cottager, and between farmer and farmer, competition tended to draw people together as members of congregations though other issues divided them. But for competition between chapels to contribute towards solidarity demanded not only that chapels should have sufficient in common that the rivalry between them should be competition and not mere opposition, but that there should be a number of chapels. The opportunity to establish a number of chapels demanded local independence as against an order in which power was in the hands of a hierarchy that was socially distant. These conditions were better met by nonconformity than by the Established Church for in the latter the source of power was remote in two senses, geographically and socially, while in both senses the nonconformist leadership was nearer to the common people.

It would appear that in Wales generally the social distance between the gentry and the common people had widened during the late eighteenth and early nineteenth centuries. The landed proprietors were enriched during the Napoleonic Wars when rents were raised as the price of agricultural products rose. At the same time the 'new agriculture' reached south-west Wales, making new financial demands on landowners. On the one hand many of the yeomen sold out so that one link between the gentry and the tenant farmers declined in importance.[28] On the other hand during this period the gentry undertook the rebuilding of their mansions in more elaborate style or built new mansions (as was the case at Gernos mansion in Llangynllo parish and Gwernant mansion in Troedyraur parish

---

[28] It is possible to gain some indication of the decline in the numbers of the yeomen in South Cardiganshire by comparing the number of male freeholders whose names and holdings are given by J. H. Davies in his 'Cardiganshire Freeholders in 1760', *West Wales Historical Records,* vol. iii (1912–13), pp. 73–116, with the relevant information in the Tithe Schedules of *circa* 1836–40 (in N.L.W.). In Troedyraur and the adjoining parishes of Betws Ifan, Llangynllo, and Penbryn, there were forty male owner occupiers in 1760 and seventeen male owner occupiers with holdings of thirty acres or more when the Tithe Schedules were prepared *circa* 1836–40.

respectively). The same years saw the first granting of annual leases of agricultural land instead of leases for life or for lives, or for extended periods of years. In 1811 the breach between the Calvinistic Methodists and the Church became complete when the Methodists for the first time ordained their own ministers, with the result that many gentry families who until then had regarded Methodism favourably became less well disposed towards it. It was from this period on that there was a rapid growth in the numerical strength of nonconformity and an elaboration of that competition between chapels which contributed to the internal cohesion of the individual congregations.

This section may be closed with a word of warning. The above does not of itself explain the strength of nonconformity or the number of chapels in South Cardiganshire. It constitutes part of the sociological context in which individuals lived, influenced by many other considerations than those which have been here discussed.

* * *

In the idiomatic expressions of their speech people recognized a distinction between those who were members of religious bodies and those who were not. On the one hand there were the 'people of the chapel', and on the other 'the people of the world'. This distinction was understood to apply to formal membership of, and not simply to attendance at, places of worship. Some who were 'of the world' were regular attenders at places of worship, and 'hearers' of the Word, without being members.

It was not the practice for parents 'of the world' to have their children baptized as infants, and should these children later seek membership of a chapel their names were first placed on the chapel's books on 'On Trial', before being received to full membership 'from the world' (o'r byd).[29] Members' children were baptized at birth (except for those of the Baptist persuasion), and at the age of ten their names were placed on the chapel's books not as 'On Trial' but under the heading 'Baptized'. At ages ranging from twelve to sixteen, but usually fourteen to fifteen, they were received into full communion. This happened when there was a sufficient number to constitute a class, in which they received instruction prior to admittance to full membership 'from the church's seed'.

Certain of the chapel's services were open only to members—

[29] In this and in what follows there were certain variations between chapels but these are noted only if they were of particular significance.

communion and the 'society meeting'. 'Hearers' could and did attend the preaching services and prayer meetings. It was at an 'acceptance meeting' (*seiet dderbyn*) that those 'of the world' presented themselves when seeking membership. There they were questioned closely about their Biblical knowledge, and especially about the life and suffering of Christ, in order to see what knowledge they had of the meaning and significance of the sacraments for it was on being received into membership that they became communicants. They were baptized in the same meeting and they were accepted or rejected by a show of hands of the members present, their neighbours who knew them in their daily lives and who were familiar with their conduct. Some were placed for a while 'On Trial', for periods that might exceed three months, but not longer than six months.

Before those who were accepted made a public confession of their faith a charge was delivered to them of what was henceforth expected of them. In this two subjects were stressed—the duty of conforming with established religious practices, and the moral character that was expected of a member. Conforming with the established religious practices involved supporting all the work of a chapel by a faithful attendance at services, particularly the communion service, and the observance of Sunday as the Lord's day, a day when no unnecessary work should be done, nor conduct indulged in which was seemly enough on other days of the week. Coal and firewood were drawn and water carried to the house on the Saturday. The men shaved on Saturday evening. Sunday work on farms was limited to the minimum that was necessary in tending the animals, in fact limited to the same work that was done during the days between the death and burial of a member of a farm family.

Attendance at services and Sunday observance were connected with the second chief subject of the charge delivered to new members—the moral character expected of them. One of the chief concerns of religious bodies was the development of a particular type of ideal character and what was stressed in nonconformist chapels was the importance of sobriety, sexual morality, pure speech, honesty, and these not simply as individual virtues but as components of the character of an upright man. The practice of these virtues was 'to rear character' (*magu cymeriad*); failure to practise them was to 'lose character' (*colli cymeriad*). These phrases implied a comprehensive judgement.

Those who were to be received 'from the church's seed' attended weekly classes for two or three months before being received into communion, where they were instructed by the deacons and the minister, and were received into membership on a Communion Sunday after the sermon had been delivered. They then walked forward to the deacon's pew, and were joined by any who had been received 'from the world' but had not yet taken communion. There they were each charged personally and by name, each promised to support the various services and to observe the character expected of him, and each received the hand of fellowship from the minister. Thereupon they walked back to their seats and partook of the communion that followed.

In contrast to 'people of the world' members had been received into communion, they had made a public profession of their Christian faith, had accepted obligations to support the chapel's activities and to cultivate a certain moral character, they had also accepted the other members' right to discipline them if they were in breach of their obligations. Discipline might be restricted to a reprimand in the presence of other members, it might involve the withholding of communion until the member was either restored to communion or expelled from membership. It might lead to immediate expulsion, though the person might later seek re-admission to membership. All decisions to expel and to readmit were taken by a show of hands by kinsmen and neighbours who were also members of the same chapel. In a Calvinistic Methodist chapel in the area with which this study is concerned during the years 1880–1920 three were suspended from communion to be admitted again within the year, and twenty-six were expelled from member-ship. The two offences that most commonly led to expulsion were drunkenness and unchastity, and not uncommonly those who were disciplined and their nearest kin considered that there were others as deserving of punishment but who went unreproved. People who were not members had not accepted these obligations, were not liable to such discipline, had no part in disciplining others. They had no admission to communion. They did not attend meetings of the 'society', and during a 'communion meeting' they left when the sermon was over, before the communion service itself began.

But despite these differences in the duties and the obligations of members and non-members, the distinction between the Church and the 'World' was by no means clear except in a formal sense, and that for two reasons. On the one hand the vast majority of adults were members, and on the other some who were not members were nevertheless prominent in the affairs of some chapels. Among 'the people of the world' there were cottagers, shopkeepers, and substantial farmers, and many of the non-members were not only regular 'hearers' at those meetings open to them, but Sunday school teachers and in some cases precentors. It has previously been remarked that membership of a chapel and of its Sunday school were formally distinct, one might be a member of either without being a member of the other. When first established the Sunday school was regarded as a secular activity undertaken by a chapel, and in some areas it was objected to on the grounds that it desecrated the Lord's day. Singing classes were quickly established in the area as from the early 1860s with the introduction of the Revd. John Curwen's system of Tonic Sol-fa. These classes were frequently connected with the Sunday schools, for singing the anthem at the festivals that brought several chapels together was specifically the work of the Sunday schools. By the end of the nineteenth century singing classes followed the evening service throughout the winter, but there are known cases of people who as young men during the 1860s were reprimanded by the deacons for desecrating the Lord's day by attending singing classes (where only hymns were taught) on Sunday nights. It is known that in later years some of the very same people became esteemed workers in their chapels because of the contribution they made to the singing in those chapels. When activities that were in the first place additional to the main work of the chapels were included as part of their main work, 'hearers' who were connected with such work became prominent in the activities of some of the chapels.

Outside the chapels neither distinctions of denomination nor distinctions between members and non-members affected the composition of groups of farmers co-operating in agricultural activities, nor the composition of the work groups of individual farms. In one respect only does the membership and practice of these groups seem to have been influenced by the views and practices that have been described above. The only time of the year when beer was available on farms was during the hay harvest. Ill feeling and scuffles between those who had drunk too much were well-known features of these occasions. Teetotal farmers were unwilling to join in the work at farms where beer was provided and various farmers who

were not themselves teetotallers forwent their practice of providing
beer in deference to those who were teetotallers on principle. Teetotalism was a major issue in this area. It led to some individuals
leaving the chapels for the Church, and it led to a complete split
in the congregation of one chapel. But after the revival of 1859 had
drawn the overwhelming majority of the population into membership of one religious body or another, the distinction between the
'Church' and the 'World' was not to be clearly perceived. This was
the condition of drawing as many people as possible into the
activities of the religious bodies.

*   *   *

The gentry drove to church in their carriages attended by their
servants, who waited outside until their masters had entered and
then followed them to sit in the rear. It was the practice for nonconformists to walk to chapel and only the ailing went there by pony
and trap. The overt values discouraged display. People walked in
the company with whom they happened to fall in, farmers and
cottagers alike. The young men, farmers' sons and servants, sorted
themselves out from their elders and walked together in groups, as
did the young women. But on entering the chapel they reconstituted
themselves into families, sitting in the family pews. In practice as
has been seen, different members of the same family might belong
to different places of worship, and the pews were frequently known
not by the name of a family but by the name of the dwelling-place
of those who occupied it or by the name of the family home of
married siblings who had remained members of the chapel where
they had been 'reared', as the expression has it.

The members had pews on the ground floor of the chapel; those
who were not members of that particular chapel (whether 'of the
world' or people who were members elsewhere) sat in the gallery
pews. Some large farms had two pews, one on the ground floor for
the family and another on the gallery for the servants (who might
well be strangers to the area). A farmer's children sat with the
parents in the 'family' pew, the daughters invariably, though on
occasion the sons might sit with the servants in the farm's gallery
pew. In some cases too a respected 'hearer', though 'of the world',
had a pew on the ground floor. But there was a strong feeling among
the members that it was 'their' chapel, and the seating reflected the
division between the members of that particular chapel on the

ground floor, and the other attenders who sat in the gallery pews. On leaving the chapel those who had constituted themselves into family groups or sibling groups broke up, with the young men, both farmers' sons and servants together, the young women in another group, and their elders taking the opportunity to talk to friends and relatives before dispersing to their homes.

It was on the occasion of death and burial, above all else, that customary chapel practices gave expression to the groups that existed in the community. During the days between death and burial the deceased lay on a board in the parlour. While practice varied in different localities, it was common to cover the corpse with a black cloth if he or she had been married, with a white canvas if unmarried. The 'near kin' and affines of the deceased constituted and were known as 'the family of grief' (*teulu galar*). This usually consisted of the parents, grandparents, children, uncles, aunts, siblings, first cousins, nephews, nieces, of the deceased, and their spouses, as the case might be, though its composition was not rigidly defined. In the funeral procession the 'family of grief' walked as a unit with kin taking precedence over affines. A distinct gap separated them from other mourners. During the service in the chapel the 'family of grief' sat as one distinct unit. In some chapels it was customary for all other mourners to occupy the gallery pews, with the ground floor empty but for the 'family of grief', however small its number, which sat centrally, clearly distinguished from all others present. In other chapels the 'family of grief' occupied the central pews of the ground floor, surrounding pews being left vacant, so that again the 'family' was physically separate from the rest of the congregation. In either case the members of the 'family' remained seated throughout the service.

On proceeding to the graveside the 'family of grief' followed the coffin, with a clear gap between its members and other mourners, which was maintained while the body was committed to the grave. The 'family' then filed past, to be followed by the other mourners. The Sunday following the funeral, the members of the 'family of grief', wherever they were themselves members, attended at the chapel where the burial service had been conducted, sitting as one unit, and remaining seated throughout the service. Thereafter it was only the immediate family of the deceased who remained seated throughout each service. This they did for six months, while mourning was worn for a full year from the date of the burial. Where families were too poor to be able to afford mourning for all members, it was worn by the surviving head of the family.

It has been noted that the minister's charge on receiving people into membership stressed the ideal character towards which members should strive. The notion of 'character' had a most prominent place in the teaching of the churches and guided their practice. The offences for which people were disciplined were either failure to attend 'society' and communion services, or shortcomings that led to a 'loss of character', to becoming a 'weak character'. The ideal character as conceived by the nonconformist churches was relevant to all members regardless of their worldly station, and the behaviour demanded, sober, industrious, thrifty, was such that would enable a man to develop his character, and direct that development. One of the reasons for the importance that the churches attributed to family life was that the development of the children's character in conformity with this ideal was a primary duty of all parents.

In this connection it needs to be noted that this was conceived of as a specifically 'nonconformist character' which in certain respects was seen as contrasting with the lives of the gentry and constituting a different ideal at which to aim. The local nonconformist leaders were convinced teetotallers whereas the gentry were not. The ministers were anti-militaristic whereas the gentry families had close connections with the Armed Forces. The nonconformist leaders did not see the life of the gentry as the pattern that others ought to emulate. Among the duties that the nonconformist leaders laid on the members of their churches was that of self-improvement through education. It was a man's duty to develop his talents, both for the cultivation of his own character, and in the service of his fellow members. All the ministers in the parish of Troedyraur were educators, teaching in Welsh, while none of the gentry had more than a passing knowledge of the language. The Revd. David Oliver who became the Calvinistic Methodist minister at Twrgwyn (in the village of Rhydlewis) in 1867, started teaching young adults the following year, having essays submitted to him weekly (15 on 6 October, 13 on 13 October, 18 on 21 October, 1868). With his helpers he continued the work for almost forty years. The Congregational ministers undertook similar work, including conducting classes in Welsh grammar using Dewi Mon's (the Revd. David Rowlands, 1836–1907) 'Grammar' as a textbook. This was the *only* instruction in writing the language that people received unless in their own homes, for elementary school education was in English. The ministers themselves were distinguished. The Revd. David Adams, B.A. (Congregational) was a nationally known literary and eisteddfodic figure; the Revd. E. Keri Evans, M.A. (Congregational), had held the chair of Philosophy at the University College of North Wales, Bangor.[30] The gentry were both members of an empire-wide social stratum whose cultural tradition was English and of a local community whose literary culture was Welsh: ministers and prominent laymen of the nonconformist churches were members of Welsh nation-wide denominational networks and of a Welsh local community wherein effective leadership fell into their hands. The close connection between the influence of the religious bodies and a high evaluation of education is seen in this: of the children living in the area of one Board school in the parish of Troedyraur, sixty-six received a secondary or higher education during the years 1890–1912; thirty-two of these were the children of ministers and deacons, who were numerically a very small proportion of the adult population.[31]

The importance attached to education was of particular consequence in this rural community in which farmers' sons generally sought to be farmers themselves in a period when there were too few farms for all of them. It contributed to evaluating as success other occupations than farming and opened the way to those occupations. One process of social change which was at work was that

[30] The Revd. David Adams was one of the first graduates of the University College of Wales, Aberystwyth. He formulated his ideas early in his career, publishing an essay entitled 'Creed and Character' in the first issue of the college magazine (Aberystwyth, 1878). For his views on war and militarism see his article 'Y Wladwriaeth a'r Bregeth ar y Mynydd', *Y Dysgedydd* (Dolgellau, 1916), pp. 156–61. In their biography of the Revd. David Adams, E. K. Evans and W. P. Huws quote part of a poem, 'Education', with which Adams won the chair at the Powys Eisteddfod in 1877;

> *Y garw farmor roddir i ni yn ddefnydd*
> *Ond Addysg sydd i'w gynio'n ddelw ysblennydd*
> *Rhydd Nef yr hedyn—ni sydd i'w ddatblygu,*
> *Ni thyf planhigion enaid heb eu garddu.*

(E. K. Evans and W. P. Huws, *Cofiant y Parch David Adams* (Liverpool, 1924), p. 25). This may be rendered: 'It is rough marble that is given us for our substance but it is Education that is to chisel it into a splendid form; Heaven gives the seed —we are to develop it, the soul's plants grow not unless they are cultivated.'

[31] Sources: Cardiganshire Local Education Authority Records; school log books and registers (now in N.L.W.), and personal inquiry. The registers themselves do not provide a complete record as some who started work on leaving elementary school entered private grammar schools later, when they had saved sufficient money.

whereby farmers sought to establish fewer of their children on the land and more of them in other occupations. In this connection the emphasis laid by religious bodies on self-improvement and self-enlightenment gave a particular direction to a change in values and thereby affected the character of the general change that was occurring in the community. For education and professional advancement were virtually synonymous with emigration, and the loss of the educated to the locality.

IX

THE RELIGIOUS REVIVAL OF 1904–1905

WALES experienced a religious revival during 1904 and 1905, the latest of a number of revivals that have affected the country since those of the eighteenth century. In certain regards this last revival was closely connected with South Cardiganshire though its leading figure was a young man from Loughor in West Glamorgan, and it affected the population of the rural areas and the industrial towns alike. While this chapter is not concerned with the 'origins' of the revival, it will be necessary to describe events connected with the revival and then to note certain of its sociological features and consequences.

Much of what has been said so far in this study has emphasized the 'parochialism' of the community which has been described, but this is not to be understood to mean that the area was isolated from events in the country at large. During the second half of the nineteenth century there was in Britain a great deal of church missionary work among those who were without the fold of the churches as well as revivalistic work among those who were within the fold. The Salvation Army was established in 1865. In 1875 the first Keswick convention was held, a convention for practising churchmen as contrasted with missions to those who were outside the church. The first of the university settlements, Toynbee Hall, was founded in 1884. In 1891 Dr. John Pugh (1846–1907), a Calvinistic Methodist minister, began an evangelical campaign in a tent in Cardiff and from this grew the Forward Movement, a home missionary establishment with which the brothers Seth and Frank Joshua were soon connected, the one best known as a revivalist and the other as a 'gospel singer'. In 1893 the Revd. John Evans (1840–97) of Eglwys-bach, a Wesleyan minister, was 'given freedom from the ordinary plan of his church' to found a mission in Pontypridd. He had visited the United States twice, in 1873 (the year in which Moody and Sankey began their first mission to Britain) and in 1887. It may be recalled that it was after the return of the Revd. Humphrey Jones from America that the revival of 1859 began in

North Cardiganshire. During these years there was a widespread belief in the efficacy of prayer and among many a conviction that God did 'touch' men. There was a growing concern for the physical and moral state of the poor whose plight was most obvious in the industrial areas and the great cities. But there was also a conviction among many clerics and laymen that religion was not to be equated with morality or humanitarianism. They regarded these virtues as the outcome of religious awareness not as its equivalent.

In this account of the background to the revival of 1904 and 1905 it will be seen that the revival was directly linked with a series of conventions, the first of which was held at New Quay (Cardiganshire) in the last days of December 1903 and the first days of January 1904, that those who were invited to address this convention were people who attended the Keswick conventions or their offshoots and that these same people were again among those who had come under the influence of a minister lately returned from the United States to minister to a church in the town of Carmarthen. For this reason the Keswick conventions shall be mentioned first, the activities of a circle of ministers centring upon Carmarthen will then be discussed, and lastly events in South Cardiganshire can be described.

In 1873 an American, Robert Pearsall Smith (the father of Logan Pearsall Smith), a glass manufacturer from Philadelphia, arrived in England to summon people to 'the higher Christian life'. He worked among the relatively well-to-do, and his wife became friendly with Georgina, the wife of William Cowper-Temple (later Lord Mount-Temple) of Broadlands Park, Hampshire.[1] When Robert Pearsall Smith suggested holding a convention for Cambridge University students on the model of American camp meetings, Cowper-Temple offered to put Broadlands Park at his disposal. The convention was held in July of 1874 and it was decided to hold a further convention at Oxford.

This was held, and to it went Canon Thomas Dundas Harford-Battersby, Vicar of St. John's, Keswick. On the final evening of the conference he expressed his own experience there: 'I feel most thankful to have shared in this Pentecostal season. . . . It is difficult to speak of my own experience, and very distasteful, yet perhaps for this reason it may be right to do so and to acknowledge the

---

[1] W. Cowper-Temple was the author of the Cowper-Temple clause in the Education Act of 1870, whereby unsectarian religious teaching was provided in the 'Board' schools.

blessing I have received.'[2] Canon Harford-Battersby also went to a convention that Robert Pearsall Smith held at Brighton in May 1875, and soon after he invited Pearsall Smith to address a conference that he (Harford-Battersby) would arrange at Keswick. However, when the arrangements were completed but before the convention met Robert Pearsall Smith left Britain, and the first Keswick convention opened without him on 28 June 1875.

Conventions were afterwards held annually at Keswick and a number of Welsh clergy and laymen came to attend them. In 1896 thirteen of them had met one day (during the convention) 'to pray that God would give Wales a Convention for the deepening of spiritual life';[3] and it was said of them 'that from that time on, they had been holding this petition before the Lord'.[4] Two ministers recalled this at the Keswick convention of 1902. The one was the Revd. John Rhys Davies (1858–1926), a native of New Quay, who was first a Calvinistic Methodist minister but was in 1902 a Baptist minister in Southport. The other was the Revd. D. Wynne Evans (1861–1927), a native of Llanybydder and in 1902 a Congregationalist minister in Chester.[5] They approached Mrs. Penn-Lewis (1861–1927) who had addressed the Ladies' Meeting during the convention. She was the daughter of a Neath civil engineer and the granddaughter of a Calvinistic Methodist minister. She was married to the city accountant of Leicester, and while living in Surrey had been a parishioner of the Revd. Evan Hopkin (a Cornishman) who had been prominent at the Keswick conventions almost from their inception. A wealthy and evangelistic woman she was known for her interest in missionary work at home and abroad and had visited missionary fields overseas. Rhys Davies and Wynne Evans told her that ever since 1896 they had desired and had prayed for a similar convention in Wales.

During the summer of 1902 Mrs. Penn-Lewis consulted the Revd. David Howells (Llawdden, 1831–1903), an evangelical clergyman who was the Dean of St. David's. He encouraged her to proceed to establish a convention and 'he entered into all the detailed arrangements for the Conference'.[6] At Llandrindod Wells one H. D. Phillips

[2] J. C. Pollock, *The Keswick Story: The Authorized History of the Keswick Convention* (London, 1964), p. 29.
[3] M. N. Garrard, *Mrs. Penn-Lewis, A Memoir* (London, 1930), p. 221.
[4] Ibid., p. 221.
[5] D. Wynne Evans, *Yr Ysbryd Glan a Diweddar-Wlaw y Diwygiad* (Chester, 1906). This publication contains a statement of many of his views.
[6] J. Penn-Lewis, *The Awakening in Wales* (London, 1905), p. 29.

had recently organized a series of meetings for the Revd. Dr. F. B. Meyer, a celebrated preacher, a future President of the Free Church Council and of the Baptist Union, and a prominent figure at Keswick. Phillips accepted the local secretaryship of the proposed 'Keswick in Wales' convention, and this was duly held at Llandrindod Wells in August of 1903. During that convention the chair was occupied by Albert Head, one of the leaders at Keswick, and others who were very prominent at Keswick, including Evan Hopkin, Dr. F. B. Meyer, Stuart Holden, and Mrs. Penn-Lewis, appeared on the platform.

Meanwhile in 1897 the Revd. W. S. Jones had been installed as the minister of Penuel, a Baptist church in Carmarthen town. He had been born in Rhymney, the son of a stonemason, and ordained to the Baptist ministry in 1885. He achieved prominence as a preacher while ministering to churches in Pontypridd and Swansea before emigrating to become the pastor of a church in Scranton, Pennsylvania, in 1892. There 'A day came when the Reality of God as a Person and a Holy Presence seemed to dawn on his soul'.[7] After returning to Carmarthen his preaching influenced others with whom he came in contact both in the locality and through the organization of the Baptist churches. These included the Revd. W. W. Lewis (1856–1938), who was then the minister of Zion, the English Calvinistic Methodist church in Carmarthen, the Revd. E. Keri Evans, a Congregationalist Minister, who had removed from South Cardiganshire to Carmarthen in 1900, the Revd. Owen M. Owen (1871–1952), the minister of the Baptist church of Elim, Penydarren, from 1901 (who became a secretary of the Llandrindod Wells convention for many years), and the Revd. Rhys Bevan Jones (d. 1933), a Baptist minister at Porth, where he was to establish the South Wales Bible Training Institute wherein those awakened by the revival might make themselves better fitted to be of service in their own churches. Therein the Revd. W. S. Jones was first his assistant (as was Owen M. Owen) and then his successor.

W. S. Jones's preaching

revealed an ideal of Christian living far transcending the level of dead morality that ordinarily satisfies even those who profess regeneration. It insisted that the laws of Christ had, for the believer, superseded the laws of Moses. 'Be ye holy' had substituted 'Be ye moral'. In its light it

---

[7] R. B. Jones, *Rent Heavens* (London, 1931), p. 26.
[8] Ibid., p. 29.

was seen that to hate, despise, be unforgiving, etc., were as vile, if not viler, sins than even the 'gross sins' that had always been abhored.[8]

Certainly since the eighteenth century one of the most prominent strains in Welsh revivalism was the rejection of an equation of morality with religion.[9]

Interdenominational prayer meetings were started in Carmarthen during 1902 under the auspices of the Pentecostal League, which was under the presidency of Mr. Reader Harris, Q.C., who before his conversion had been one of Charles Bradlaugh's chief assistants. The following year Reader Harris and two of his assistants came to hold a convention in Carmarthen. A few months later another convention was held at Heol Awst Congregational chapel, which was addressed by the Revd. R. B. Jones and Mrs. Penn-Lewis. Its purposes was to 'deepen the spiritual life of the churches'.[10]

By May of 1903 a group of ministers influenced by W. S. Jones's preaching began to meet regularly in Carmarthen to pray and to discuss their religious experiences with one another. These wrote to Dr. F. B. Meyer asking him to come to meet and advise them. Meyer replied that he could not do so, but advised them to attend the convention that was to be held at Llandrindod Wells where he could meet them privately. Six of them went and there 'a new vision dawned on their souls'.[11] As one of them later said 'a "new world" had opened to them and they could not but lead others in'.[12] With

[9] The following hymn was current orally in South Cardiganshire during the nineteenth and the early twentieth centuries. It shows, in combination, the acceptance of certain theological views which were put forward by the leading revivalists of the eighteenth century and the down-to-earth quality characteristic of the work of many rural hymn writers. (It can be readily interpreted in terms of 'great tradition' and 'little tradition', and the contact between them, as presented in Chapter I.)

> Beth dal byw yng ngwlad Efengyl
> Beth dal crefydd oddi fa's,
> Nid yw moesau ddim yn ddigon
> Heb gael cadwedigol ras;
> Gellir cyrraedd doniau helaeth,
> Deall y 'Sgrythur lân bob darn,
> A bod 'nol yn wyneb Iesu'n
> Casglu parddu ddydd y farn.

This may be rendered : what is the good of living in the land of the Gospel, what is the good of a religion which is external to oneself, morals are insufficient without saving grace; it is possible to obtain liberal capacities and to understand every part of Holy Scripture and yet to be aback when faced with Christ, engaged in collecting the soot on Judgement Day.

The hymn was recorded by D. P. Williams (Brythonydd) and is to be found in N.L.W., MS. 15661c.

[10] E. K. Evans, *Fy Mhererindod Ysbrydol* (Liverpool, 1962), p. 49.

[11] R. B. Jones, p. 32.

[12] J. Penn-Lewis, op. cit., p. 32.

W. S. Jones they felt themselves called on to undertake revival work within the churches and the early form of the revival was that of the convention, 'the message and its appeal were almost exclusively to those within the church. The call was to holiness.'[13] Conventions were held at Cwm-bach, Dowlais, Llwynypia (whither W. S. Jones moved in 1904), Penydarren, Porth, Cefnmawr, Cwmafon, Pen-coed,

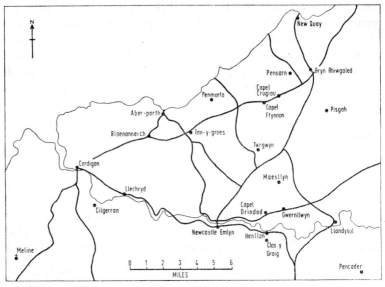

Fig. 14. Map showing place-names referred to in connection with the Religious Revival of 1904–5.

and at other places. This work started with awakenings in the churches of those ministers who had been inspired at Llandrindod Wells, and who continued to meet regularly, as one of them stated to 'spend a quiet day with God. Our meetings have been indescribable, and we have had a number of Pentecosts.'[14]

In 1892 the Revd. Joseph Jenkins (1859–1929) had been installed minister of the Calvinistic Methodist church in New Quay, and two years later his nephew the Revd. John Thickens (1865–1952) became the minister of the church of the same denomination at Aberaeron some nine miles distant. Joseph Jenkins was a native of Cwmystwyth but had removed to Pentre Rhondda in 1873 on being

13 R. B. Jones, p. 33.
14 J. Penn-Lewis, op. cit., p. 32.

apprenticed to a draper. He became the minister of Twyn Church, Caerphilly, in 1885, though he was not ordained until 1887. While at Caerphilly he conducted open-air preaching meetings and sought and received the help of Dr. John Pugh (who was to found the Calvinistic Methodist Church's home mission, the Forward Movement).[15] John Thickens was born in North Cardiganshire too, but on the death of his father he removed to the Rhondda Valley with the rest of the family. After ministering to a church in Dowlais for some two years he was installed in Aberaeron in 1894.

Though they were uncle and nephew the one was but a few years older than the other, and the connection between them was very close. They were both concerned with and disturbed about their own spiritual state and that of their churches.

They met frequently, not so much to discuss the topics of the day as to discuss themselves, and sit in judgement on their ministerial work which they considered valueless and ineffective. . . . 'Our little worlds were dark; one evening we agreed that we were two fiends because of our persistent unfaithfulness to the Crucified, and we nearly determined to flee from the presence of the Lord to some Tarsus because of the darkness of our condition.' He (Joseph Jenkins) received occasional visions of the glory of the Gospel, and his soul bowed in humility, but the mists returned again. We knew very little but of the dark night of the soul. We both longed for the persisting light of day, and our longing was our capital.[16]

Jenkins connected with one another his state of spiritual darkness and his practice of delivering 'ethical' sermons in which he condemned the sinfulness of his congregation. He defended his preaching as necessary but at the same time found it inadequate, and he spent hours on end in self-examination and prayer. Periods of inspiration came to him, as at a monthly meeting of his denomination's presbytery at Pen-sarn in October 1903.

We do not remember what he said in his address, and as we listened to him speaking of 'the grace that is in Christ Jesus' we did not rightly know whether we stood on our feet or on our heads. There came torrents of speech. There came thunderbolt after thunderbolt; we were astonished and afrightened, afrightened and astonished. We treasured

[15] For the Revd. Dr. John Pugh see *The Dictionary of Welsh Biography down to 1940* (London, 1959). For the Revds. Joseph Jenkins and John Thickens see W. Morris (ed.), *Deg o Enwogion*, Second Series (Caernarvon, 1965).

[16] E. Howells, 'Torriad y Wawr yn Ne Aberteifi', *Cyfrol Goffa Diwygiad 1904-1905* (eds. S. Evans and G. M. Roberts), (Caernarvon, 1954), pp. 28-9. Howells, himself one of the revivalists, had access to the papers of the Revd. John Thickens and is here quoting from them.

the 'grace', and our souls bowed down . . . A strange twenty minutes. The speaker sat. We have not forgotten the silence that followed. The one looked at the other as if asking 'Where have we been? Where are we?' No hymn was sung. The Apostolic Blessing was pronounced, and in the deafening sound of the silence the congregation dispersed slowly and as if in a dream.[17]

Jenkins and Thickens were concerned in particular with the condition of the 'young people' in their churches, not only because they wished to secure these in their faith while still young, but because they were both aware that a generation was arising which did not know a visitation of the spirit. They feared that people would come to look to 'the County Council, the District Council, the Parish Council, the Intermediate Schools which were new among us, for social deliverance; and that we would fall into lukewarmness in the work of the Lord'.[18] The formalism of the services, the speeches which passed for personal testimonies in meetings of the religious society, the readiness of congregations to equate morality with religion, they regarded as manifestations of the absence of a divine inspiration and of a lack of grace within.

In the monthly meeting of the South Cardiganshire presbytery of the Calvinistic Methodist Church of July 1903, which met at Llechryd, Joseph Jenkins announced that he intended to present a motion 'That means are to be adopted to foster faithfulness to the (Calvinistic Methodist) Association and to Christ'.[19] This he formally did at the monthly meeting held at Abermeurig in October of that year, and a committee of ministers was appointed to consider how to further that end. This committee, with John Thickens as its convener, met at Lampeter and the members had considerable difficulty in understanding what precisely Joseph Jenkins had in mind. The older members were thinking of a

revival, similar to that they had seen in 1859, while others among us wished to devise means 'to deepen the spiritual life' of our people. It was explained that a number of young men and women met weekly in one of our churches to this end, and that many of them had already delved deep (into the things of the spirit). We agreed that we were near to losing spiritual anxiety from our churches, and that we would lose it entirely unless the churches were soon revived by the spirit of Christ; that we would rear young men and women who had not known the way

[17] E. Howells, op. cit., pp. 29–30.
[18] Ibid., p. 32.
[19] Ibid., p. 32.

to the larger life in Christ; that we would be deprived in the course of a few years of those who had experienced the great things of the Gospel.[20]

It was decided to hold a 'Convention to deepen the spiritual life of the churches'. Neither Jenkins nor Thickens had been to the 'Keswick in Wales' convention held at Llandrindod Wells the previous summer, but they certainly knew of it as either the one or the other had met one of those who frequented the conventions held at Keswick, and had met the Revd. W. W. Lewis of Carmarthen, while forty pastors of the Forward Movement had attended the Llandrindod Wells convention. Further, the people they invited to address their own convention were people who had attended the one held at Llandrindod.

The committee reported to a monthly meeting held at Ffos-y-ffin on the first and second days of December 1903, and it was agreed that the convention should be held at New Quay on 31 December 1903 and 1 January 1904, and arrangements were made to publicize it. When it met Joseph Jenkins presided at all the meetings, which were addressed by the Revd. W. W. Lewis, the Revd. J. M. Saunders, M.A., and by Mrs. Saunders. W. W. Lewis (1856–1938) was, as has been stated, the minister of Zion, a Calvinistic Methodist church in Carmarthen. J. M. Saunders (1862–1911) was the son of the Revd. David Saunders who was President of the General Assembly of the Calvinistic Methodist Church in 1877 and the nephew of the Revd. David Howell, the Dean of St. David's, until his death in 1903, whom Mrs. Penn-Lewis had consulted when discussions were in progress to establish the first of the 'Keswick in Wales' conventions. Mrs. Saunders was the daughter of R. J. Davies, Esq., of Cwrt-mawr, a Welsh-speaking nonconformist Cardiganshire squire who was chosen treasurer of the General Assembly of the Calvinistic Methodist Church in 1873. Her brother was J. H. Davies (1871–1926) who in time was chosen treasurer of the same body, and was later to be successively registrar and principal of the University College of Wales, Aberystwyth. There were family connections between them and Prebendary Hanmer W. Webb-Peploe, who had been one of the supporters of the Keswick convention from its inception in 1875. All three chosen to address the convention had been to the Llandrindod Wells convention, where Mrs. Saunders had addressed meetings of the Forward Movement that followed the main convention.[21]

[20] Ibid., p. 32.   [21] J. Penn-Lewis, op. cit., p. 35.

Four further conventions were eventually held; at Aberaeron on 30 June and 1 July 1904, at Blaenannerch in late September of 1904, at Tregaron in January of 1905, and finally at New Quay in midsummer of 1905. All the conventions were arranged by the South Cardiganshire presbytery of the Calvinistic Methodist Church; the presbytery's churches were invited to send delegates to them (and did so); the speakers who addressed them included the Revd. W. S. Jones, the Revd. E. Keri Evans, the Revd. J. M. Saunders, Mrs. J. M. Saunders, and the Forward Movement's evangelist, the Revd. Seth Joshua. These were invited because 'they believed most passionately that it was Union with Christ which was the condition of the vitality and success of every church order'.[22] It was at the third of these conventions (held at Blaenannerch in September 1904) that Evan Roberts, with whose name the revival was chiefly to be connected, first came to prominence.

In October 1903 John Thickens started conducting a weekly meeting for the young people of his chapel at Aberaeron in which they discussed the nature 'of the eternal life in Christ Jesus'. Soon after, Joseph Jenkins instituted a similar meeting for the young people of his church at New Quay, held after the Sunday morning service. Realizing that he had little to say to them beyond urging faithfulness upon them he became increasingly aware of the short-comings of his ministry. He came, according to his own testimony, to pray for hours on end each night, and on one night early in 1904 he experienced what to the end he regarded as the all-important inspiring moment of his life.[23] In one of the young people's meetings held in late February of 1904 he listened to the speeches offered as personal testimonies in the usual way, but protested that what he wanted was 'not addresses such as we receive in the society meeting, dressed out as "experience", but the true thing in a few sentences'.[24] One young woman who had already been affected by his ministry responded in a way that changed the character of the meeting, making a public protestation of her love of Christ. In subsequent meetings and in additional week-night meetings the young were awakened. News of this spread. 'The influence worked its way, growing Sunday by Sunday. Older people began to flock to the meetings, and people from the countryside travelled to the quiet

[22] Y Drysorfa (Caernarvon), Apr. 1953, p. 72.
[23] E. Howells, op. cit., pp. 34–5.
[24] Ibid., pp. 35–6.

village (New Quay), driving fifteen miles . . . to the young people's weekly meetings.'[25] Some ministers visited New Quay and invited the revivalists to visit their churches. In a short time Joseph Jenkins was taking groups of his young people to various Calvinistic Methodist churches in South Cardiganshire, and the young men and women testified to their religious experience and besought the descent of the spirit.

Joseph Jenkins believed that others of a similar spirit in other chapels wanted him to come to their aid. He and his young people were particularly welcomed at Calvinistic Methodist chapels whose ministers had been active revivalists in 1859, notably by the Revd. David Oliver in Twrgwyn (Rhydlewis) and the Revd. Evan Phillips at Bethel (Newcastle Emlyn). They were particularly welcome too at some chapels where the ministers were very young, notably at Blaenannerch and Capel Drindod. Blaenannerch (which had been the chief diffusion point of the 1859 revival in South Cardiganshire) was under the ministry of the Revd. M. P. Morgan (1876–1964) who had visited New Quay in the company of the Revd. Evan Phillips. Capel Drindod, where the Revd. Evan Phillips had been 'reared' (to employ a local expression), was ministered to by the Revd. Rhystyd R. Davies (1880–1942).[26] In Joseph Jenkins's words, 'Soon, I went and a number of the Tabernacle's youth with me, to Twrgwyn, Capel Drindod, and other places, and we sought to spread to the other churches the fire that consumed us, but we were most careful that the flesh should not rejoice'.[27]

As has been said the Revd. Evan Phillips had evangelized during the revival of 1859 (in the company of one of its main figures, the Revd. David Morgan) and welcomed the revival of 1904. His son, the Revd. John Phillips, a Calvinistic Methodist like his father, kept a private grammar school in Newcastle Emlyn. One of his students was Evan Roberts. Aged twenty-six in 1904, Roberts (1878–1951) was a native of Loughor in West Glamorgan and after having been a miner and then a blacksmith he applied successfully to his presbytery for permission to study for the ministry. When he entered the Revd. John Phillips's grammar school he had been praying for spiritual experience over a period of thirteen years. He was present at a revival meeting at Twrgwyn, where Joseph Jenkins

[25] R. R. Davies, 'Torriad y Wawr yn Sir Aberteifi', *Y Diwygiad a'r Diwygwyr* (ed. T. Francis), (Dolgellau, 1906), p. 53.
[26] E. Howells, op. cit., p. 36.
[27] Ibid.

had brought his young people to testify to their experience. There he was heard to state, 'A hundred thousand souls for Christ', which became one of the battle cries of the revival. Even those ministers present who were concerned to see the revival spread were dubious and merely inquired who the young man was.[28] He visited the third of the conventions that the South Cardiganshire presbytery had arranged, held at Blaenannerch, in late September of 1904. Listening to Seth Joshua's address he fell on his knees, his face streaming with sweat, and took up a phrase uttered by Joshua, 'Bend me, Lord'. The meeting broke out in a tumult. Later that day the young people of Blaenannerch held their own meeting in the chapel, in which were seen the same phenomena as those which characterized the meetings of the young people of New Quay. On their way home that evening some of the revival's leaders agreed that the events during Seth Joshua's meeting had diverted the convention from its proper course.[29]

Evan Roberts returned to Newcastle Emlyn and with eight others he intended to go on an evangelizing mission through Wales. He went to New Quay to discuss the matter with the Revd. Joseph Jenkins but came to no definite decision.[30] Then on hearing the Revd. Evan Phillips preach on the following Sunday at Newcastle Emlyn he decided to return to Loughor and start his evangelical campaign. This he did and in a very short time his name virtually displaced all others as the leader of the revival. The awakening spread among all denominations in South Cardiganshire, but it was Evan Roberts's work that henceforth attracted most attention. It covered the whole of Wales and lasted into the summer of 1905.

Before turning to notice certain features of the revival it should be noted that up to the time when Evan Roberts started his missionary work the leaders of the revival who had called for and initiated the conventions, who had addressed them, who had established the young people's meetings, were all either ordained ministers or laymen who were already very prominent in denominational work. In South Cardiganshire they worked through the Calvinistic Methodist Church's normal presbyterian organization, acting during the presbytery's monthly meetings which were and are only attended by ministers and deacons. But after September 1904 the leadership

---

[28] Personal information from the Revd. M. P. Morgan who was present at the meeting and heard Evan Roberts's remark.
[29] E. Howells, op. cit., p. 37.
[30] *Y Goleuad* (Caernarvon), 20 Jan. 1905.

passed to a young layman, previously little known. One of the (then) young ministers (the Revd. Eliseus Howells) who was a revivalist in 1904 and 1905 later recalled

As the Revival proceeded, Jenkins and the other brethren who had been concerned with these conventions, were worried because they thought that the Word was being dethroned, and that men were readier to speak to God rather than to listen to what God had to say to them through the preaching of the Word. . . . The leadership passed into a young man's hands, to the hands of one considered less suitable than themselves, but 'my thoughts are not your thoughts'. The Revival took a very different path from the one that they wished it to take, but 'your ways are not my ways, saith the Lord'.[31]

Before the revival began, the work of managing a chapel's affairs and of conducting its meetings was in the hands of the minister and the diaconate. Though women were the most numerous attenders at prayer meetings and at meetings of the religious society, the public work of reading, praying, contributing testimonies of one's experience, and addressing the meetings on relevant topics was exclusively the work of the men present. Apart from Sunday school work and 'young people's' activities the contribution of the women was virtually limited to reciting a few verses when asked to do so. Women who did become prominent were extremely few in number and were regarded as oddities.[32]

The young people had their own meetings which were organized by and in the charge of the minister or the diaconate, and these meetings were attended by a number of adults to guide, oversee, and help with the proceedings. In other meetings it was occasionally the case that instead of the usual scriptural reading a young man or a young woman rose to recite a chapter, but this was normally the extent of the young people's participation in meetings. Their place in the chapel was as learners, under the authority of their elders. Of the men it was the same few who took part in prayer meetings and society meetings week after week and then only at the behest of the minister or, in his absence, of the deacon conducting the meeting. This was part of the formalism against which Joseph Jenkins protested, which one of the revivalists described as an 'unhealthy procedure', and which seemed to the revivalists the manifestation of a lack of awareness in the congregation.[33] One of

---

[31] E. Howells, op. cit., pp. 37–8.
[32] Y Geninen, 1909, p. 157.
[33] T. Francis, 'Y Diwygiad yng Nghasllwchwr a Gorseinion', Y Diwygiad a'r Diwygwyr (ed. T. Francis), (Dolgellau, 1906), p. 125.

the features of the revival was that it upset or reversed the roles of different members of the congregation: the young people commonly took over the meetings, women became extremely prominent, revival meetings were notable for their spontaneity and lack of formal procedures, and the revival itself came to be connected largely with the names of laymen rather than the ministers.

It was at the young people's meeting that the revival began in New Quay and it was groups of these young people that Joseph Jenkins took with him to testify at other churches.[34] These were described in contemporary press accounts as 'bands'. Bands of young people from Tabernacle Calvinistic Methodist church, New Quay, visited Calvinistic Methodist churches at Blaenannerch, Penmorfa, Cardigan, Cilgerran, Twrgwyn, Capel Drindod, Newcastle Emlyn, which are in South Cardiganshire and the Teifi valley, and Aberystwyth in the north of the county. Those from Towyn Congregational Church, New Quay, visited Capel Crugiau, Gwernllwyn, Pisgah. Young people from Cardigan visited Aber-porth and Blaen-cefn, young people from Newcastle Emlyn and Blaenannerch visited Cenarth. Capel Drindod's young people visited Gwernllwyn. It was in the same way that the revival of 1859 had spread in South Cardiganshire when young people from Blaenannerch and Tan-y-groes visited Llechryd. There the revival began in a meeting where the visiting young men occupied the central part of the ground floor of Llechryd chapel, and the visiting young women lined the gallery.[35]

It was at a young people's meeting that the revival started in Caer-farchell (Pembs.) and it was at a meeting of the Band of Hope that it began in Cilgerran (Pembs.); the young people were notably prominent at Henllan, Penmorfa, Brynrhiwgaled, in South Cardiganshire, and in North Cardiganshire as well.[36] The Calvinistic Methodist denominational newspaper noted that 'Certainly, it is the young people who had the honour of hearing the voice and opening the door that the Head of the Church might enter; let them henceforth have more opportunity of fulfilling their task'.[37]

On occasion the young people took over revival meetings completely. At New Quay

The young people, men and women, have been aroused, and praying

---

[34] E. Howells, op. cit., p. 36.
[35] J. J. Morgan, *Hanes Dafydd Morgan Ysbyty a Diwygiad 59* (Mold, 1906), pp. 191 et seq.
[36] *Y Goleuad* (Caernarvon), 16 Dec. 1904.
[37] Ibid., 30 Dec. 1904.

has become a simple and acceptable task to them. We have had some incomparable meetings here during the past month, the old people as if thrust aside, and the young people taking complete control of our prayer meetings, men and women; the young women taking a public part in our prayer meetings; the young men crowding to the prayer meetings . . . the meeting is packed with people, with no one feeling any hurry to leave; the spirit of God having, as it were, filled the place, having descended on all present . . . Another aspect of things is that the young men and women repair as a crowd virtually every night to this house and that house wherever there are old ailing people, and conduct a prayer meeting for them.[38]

Before his return to Loughor Evan Roberts had been present at meetings which had been attended by the young people of New Quay and their minister the Revd. Joseph Jenkins, at Newcastle Emlyn, Blaenannerch, Twrgwyn, and Capel Drindod. During the month between his experience at Blaenannerch and his return to his home he tried to hold young people's meetings whenever and wherever he preached in his capacity as a ministerial student. When he did eventually return to Loughor he arrived there with the intention of evangelizing among the young people of Moriah, the Calvinistic Methodist chapel in Loughor. Having started to do so on the Monday evening following his return to Loughor his next step was to arrange for the young people of Libanus chapel (Calvinistic Methodist), Gorseinon, to visit Moriah. It was during these meetings that he first enunciated what were soon considered to be the 'articles of faith' of the revival: a confession of all sins not previously confessed to Christ, dissociation from all doubtful pursuits, a public profession of Christ, and a ready obedience to the promptings of the Spirit. By the end of the week it had become clear that the awakening would not be confined to the young, but as the revival spread it was young people, men and women, who came to prominence as its leaders.[39]

The women took virtually no individual part in meetings before the revival, nor did they for many years afterwards, but during the revival they became active and prominent in prayer meetings, as members of the bands of people visiting other chapels, and as small groups accompanying evangelists to their meetings. Many became national figures, known for their testimonies and as 'gospel singers'. At Caerfarchell it was the women who were the first to join the

[38] Ibid., 22 Apr. 1904.
[39] T. Francis, op. cit., pp. 57 et seq.

young people in the Saturday night prayer meetings wherein the revival started in that locality. At Cardigan the women established their own prayer meetings, and these 'increased Sunday by Sunday without being announced, it being an entirely spontaneous movement on their part'.[40] At Abermeurig 'The revival . . . has taken the form that is common in this Revival—the young women and older women have taken to praying, as well as the young men'.[41] At Brynrhiwgaled 'The young people, the young women, and the older women have received bountifully of the fire. Prayer meetings are held too by the young people themselves, and in these it is the young women who first break out in praise and prayer'.[42] Contemporary accounts repeatedly mention the part played by women, and note it as exceptional: it is, too, well remembered.

Revival meetings were arranged when revivalists were invited to pay a visit to a chapel; they were held too when the revivalists simply sent word that they were coming, and at other places people pressed the minister and deacons to arrange meetings. At some chapels six meetings were held on Sundays (including the normal services) and in places there were prayer meetings on every night of the week. They lasted several hours, well into the night. In these meetings the usual order whereby people prayed when requested to do so was abandoned and it was open to anyone to pray whenever he or she felt impelled to do so. As meetings developed numerous people prayed or sang or offered testimonies independently of one another and at the same time. Not uncommonly those who had offered prayer when called upon to do so before the revival began now desisted. Ministers found themselves unable to address the congregation. (This is the point of the remark quoted above, that 'the Word was being dethroned, and that men were readier to speak to God rather than to listen to what God had to say to them through the preaching of the Word'.) The emphasis was on spontaneous and *extempore* prayer, singing, and rendering testimonies. The relatively formal elements in worship, and to a degree the sermon, were largely displaced. The revivalists themselves including Evan Roberts offered addresses rather than formal sermons and many of their hearers became conscious of the sinfulness of many activities that they had not previously considered sinful. They

[40] *Cardigan and Tivy-Side Advertiser*, 9 Dec. 1904.
[41] *Y Goleuad* (Caernarvon), 23 Dec. 1904.
[42] *Y Tyst* (Merthyr Tudful), 25 Jan. 1905.

became aware too of their own sinful state and sought deliverance from it for themselves and their acquaintances. People prayed for themselves and for others by name and those who had not made a public profession of Christ were entreated to do so and to join the church.[43]

When prominent revivalists were to conduct meetings, work on farms in the locality ceased and on occasion the woollen mills of the Teifi valley closed down.[44] It is hardly possible to describe Evan Roberts's meetings except by quoting eye-witness accounts. There follows a description of a meeting held at Blaenannerch in mid March 1905. It is by D. P. Williams (Brythonydd), who later entered the Baptist ministry. There are several other eye-witness accounts of the same meeting.

In this meeting Aerwyn and I stood on our feet near to the deacons' pew. The chapel was packed, and we were in from the beginning to the end. Evan Roberts led this meeting throughout. After ascending to the pulpit he sat for a moment or two. The young women who had accompanied him lay down on the floor of the deacons' pew. The crowd was singing. When he rose he forbade the singing and said, 'you cannot understand why I prevent you from singing, but my reason is that it is hypocrisy, and that God does not accept your praise'.

Then he spoke for some ten or twelve minutes, fluently, and in language that was easily understood. One of the remarks he made was that there were in the past two kings among the Jews, the king of Israel and the king of Judah, but that only one king was to rule the heart of a sinner. He stopped suddenly, and then said, 'My speech has come to a stop, and when I stop, I stop. It is thus far that the spirit has provided and I do not know what in the world to say or do next'.

Then he sat and fell into a spasm, and then in some two minutes time he rose and held the Bible up, asking 'What is this?', many answered 'A Bible'; another said 'The Word of God'. 'Yes' said he, 'and it is divine, but there is one present tonight who denies it'. At once it was felt that terror walked the place. The young women who were on the floor of the deacons' pew were wailing and praying, and E(van) R(oberts) called on him (the doubter) to rise immediately and confess, that there was no time to lose, and that it was not possible for the worship to proceed while he was there. After a few moments he (Evan Roberts) said that he (the doubter) was on his feet, and that he had not attended the afternoon meetings. He called on him to confess at once lest God strike him dead, that such things had happened more than once in recent weeks when people remained obstinately and firmly opposed to God's command.

[43] For the abandonment of the normal procedure at meetings see: T. Francis, op. cit., p. 125 ; *Cambrian News* (Southern Edition), (Aberystwyth), 15 Dec. 1905.
[44] *Cambrian News* (Southern Edition), (Aberystwyth), 18 Nov. 1904.

The young women and others prayed 'Bend him Lord'. 'O! bend him Lord, for thy son's sake'. 'To thy Glory'. E(van) R(oberts) went from spasm to spasm, while announcing new and different threats. At one point he said 'If he does not confess I shall probably get his name, and once God tells me his name I shall have to announce it'. At another point he announced his (the doubter's) age: 'He is a young man of twenty-five'. Then, 'I have been given his name but I have not been commanded to announce it'. There was loud wailing and tumult. The next time that he recovered from a spasm he announced the name, 'His name' said he 'is Thomas Walters, and he is thirty three years of age'. He now called upon him authoritatively to confess at once. He fell in a spasm upon the Bible and the desk.[45]

One feature of the meetings that Evan Roberts conducted, not revealed in this account, was the attempt that he made to put an end to long-standing dissensions and disagreements between members of the chapel where a meeting was being held. At the meeting described above he referred to a quarrel between members and announced, 'There are people here who have done injury to their fellow men, and they do not live far from here. God does not call them to confess publicly, but they must make restitution to the injured persons as soon as possible'.[46] He had specific people and specific offences in mind. He rarely named offenders, but referred to them obliquely as being of a certain age, present at that meeting but not at another one, having long been a member of the chapel or having been a member for only a short time, as the case might be. Everywhere he was insistent that dissension must be removed and wrongs put right, that people must be at peace with one another before they offered prayer and sought blessing.

In March 1905 Evan Roberts proceeded from Blaenannerch to New Quay, and a description of part of the meeting that he conducted there is now quoted:

Several hymns were sung, and in the meantime the missioner was in silent prayer, and, judging by the twitching of his face and body, he appeared to be in great agony. Someone in the body of the chapel started a hymn. Two lines, however, had not been sung when Evan Roberts awoke as if from a trance, and said, 'Stop! Stop! He will not

[45] This account is to be found in N.L.W., MS. 15668B. Another account of the same meeting is to be found in Awstin, *The Religious Revival in Wales*, No. 5 (Cardiff, 1905). A further account by the Revd. T. Phillips, who was the rector of the parish of Aber-porth in which Blaenannerch chapel stands, is to be found in *Yr Haul*, vol. vii, 1905, pp. 161–4. The Revd. T. Phillips joined in revival meetings held in the Calvinistic Methodist chapel at Aber-porth, and conducted similar meetings in the parish church.

[46] *The Welsh Gazette* (Aberystwyth), 16 Mar. 1905.

accept that sort of singing. There is an obstacle here which must first be removed.'

It transpired afterwards that the obstacle was disobedience on the part of those who were church members. He asked them all to pray that the obstacle might be removed, and then followed a scene which had not before been witnessed at New Quay, the simultaneous praying and weeping of four or five hundred people. This had gone on for some time when Mr. Roberts said that the difficulty had passed off, and that they might now sing. Throughout the crowded chapel persons might be seen speaking to their neighbours, and . . . (hymns) were sung with great emotion.[47]

Roberts believed in God's 'word' not only as a source of inspiration and authority but as conferring power in a way which reminds one of the verse 'So shall my word be that goeth forth out of my mouth: it shall not return unto me void, but it shall accomplish that which I please, and it shall prosper in the thing whereto I sent it'.[48] In this context by God's 'word' Evan Roberts intended not the Scripture in general but specific words and phrases which gave their recipients power in respect of what those words and phrases referred to, and further that the words themselves possessed the power to sustain their recipients in what of God's work was indicated by those particular words and phrases. 'There is', he wrote, 'in this word life and divine power, and it can grasp you and hold you to itself.'[49] The idea that the word or phrase contained the power of He who spoke it to the recipient can be seen too in the introductory passage to the first sermon that Evan Roberts wrote, some four months before his first visit to Blaenannerch.[50]

It is probably inevitable that most people found and find a man as unusual as Evan Roberts a strange and puzzling character. An acute observer (E. Morgan Humphreys) who came to know him well, and attended many meetings addressed by him, and who, although initially sceptical of Evan Roberts, later, like many others, changed his mind about him, commented:

It was gradually that I came to know him well. We lodged in the same places and we went to the meetings in the same conveyance; we walked together and even tried to compose an *englyn* together. . . . The more I saw of him, the more my liking for him deepened as did my respect for him. It was impossible not to respect his intellectual ability; some of his

[47] Awstin (1905), p. 20.
[48] Isaiah 55 : 11.
[49] E. Roberts, *Pedair Safon o'r Bywyd Ysbrydol* (Leicester, no date), p. 7.
[50] The sermon is published in D. M. Phillips, *Evan Roberts, the Great Welsh Revivalist and his Work* (London, 1906), pp. 465-75.

addresses were brilliant, they contained no suggestion of a superficial appeal to the emotions, and it is a great mistake to think that Evan Roberts was a religious spell-binder. He was a quiet speaker, his words and his mind orderly. He frequently spoke as a young man to other young men . . . and he always spoke naturally. I never once heard a single note of *hwyl* from him; he did not shout, but he was fearfully serious on occasion, as if his nature was racked.[51]

Criticism was commonly voiced that the official leaders of the Calvinistic Methodist Church were antipathetic towards the form that the revival had taken, and one of the denomination's best-known figures took it upon himself to still this criticism. The Revd. Dr. Cynddylan Jones, who was Moderator of the General Assembly of his Church in 1901, did so by publishing a number of letters in the *Western Mail*. In a letter that bore the heading 'Welsh Methodists and the Revivalist' he welcomed the work of the young revivalists and stated that 'they need fear no censure or restriction; we view their enthusiasm with gratitude and praise'.[52] In a later letter entitled 'Danger Signals' he found it necessary to refute statements by lay revivalists that there was no longer any need for an established ministry, a view that Evan Roberts himself came to share with those lay revivalists. Dr. Cynddylan Jones rebutted this notion by suggesting what was virtually a division of labour between ministers and revivalists:

Whereas it is the work of the Revival to blast the stones in the quarry, it is the task of the settled ministry to dress them. In 1859 there was a similar explosion, between 60,000 and 100,000 converts having been made, rough stones most of them. It has taken forty years of strenuous labour on the part of the ministers to dress them and fit them in the spiritual temple. It meant teaching, correcting, guiding, shaping.[53]

In the same letter he proceeded to mention one possible danger, that Evan Roberts might fall into the hands of the professional missioners and religious showmen.

In South Cardiganshire there was virtually no open opposition to the revival among the accredited leaders of the churches, though while some ministers were enthusiastic, others were lukewarm. The one open clash between a minister and one of the revivalists occurred at Cardigan. Revivalists, ministers, and laymen laid their emphasis

---

[51] E. M. Humphreys, *Gwŷr Enwog Gynt*, Second Series (Aberystwyth, 1953), p. 104. Quoted by permission of Gwasg Gomer.
[52] *Western Mail* (Cardiff), 19 Nov. 1904.
[53] Ibid., 22 Dec. 1904.

on 'deliverance, and the certainty of salvation'.[54] Salvation was to be achieved not through good works and moral conduct nor through participation in the sacraments as such, but through the personal experience of saving grace which would henceforth 'inform', make meaningful, and render significant morality and the sacraments alike. If one experienced saving grace, one knew it, so that one knew with certainty whether or not one had received salvation.

Since the last major revival of 1859, however, a generation of preachers had grown up familiar with the testimonies to the power of grace but without personal experience of revival. This was at a time when educational opportunities were greater than in the past and the Calvinistic Methodist Church was attaching more importance than previously to the education of its ministers. There was a certain divergence between those, ministers and laymen, who emphasized the need for inspiration and those who stressed the place of education in qualifying men for the ministry. Certain of the revival's leaders were themselves well-educated men (for example the Revds. Keri Evans and J. M. Saunders) but there were others among the well-educated who attached greater importance to rational understanding than the revivalists did. The chief attack on the revival from among the ministers came from the (Congregationalist) Revd. Peter Price, at one time of Aberystwyth and then of Dowlais, who signed his critical letter to the *Western Mail* 'Peter Price (B.A. Hons.), Mental and Moral Sciences Tripos, Cambridge (late of Queens' College, Cambridge)'. In South Cardiganshire the clash between a revivalist and a minister came at Cardigan, between the Revd. Seth Joshua and the Revd. Dr. Moelwyn Hughes. Seth Joshua (1858–1925) was one of the Forward Movement's pastors. He received little formal education even in an elementary school but at a very early age he started work driving a donkey. 'I had more out of that donkey than I could get out of any College in the land. . . . I maintain that if a man knows how to handle a donkey for three and a half years he is qualified to handle anything awkward.'[55] Whether or not Seth Joshua is rightly to be considered antipathetic to education he was certainly convinced that it was inspiration and not education that fitted a man for the ministry, and in this he was in accord with Methodist revivalists from the

---

[54] See, for example, W. Nantlais Williams, 'Trem yn ôl', *Cyfrol Goffa Diwygiad 1904–1905* (eds. S. Evans and G. M. Roberts), (Caernarvon, 1954), p. 90.
[55] T. M. Rees, *Seth and Frank Joshua, The Renowned Evangelists* (Wrexham, 1926), p. 5.

eighteenth century onwards. 'By all means', said he, 'have your Art and Theological training. But you have not secured everything even then.'[56] He emphasized the difference between 'ethics' and religion as other revivalists did. 'In our intellectual pride we have been making the pulpit a kind of professor's chair from which to preach the ethics of Christianity. We were trying to reach men's natures from without. This revival has produced better ethical results in one day than we have done since the day when this false light led us astray.'[57] During the revival Seth Joshua recalled,

Three seasons in turn I saw the white thorn blossom and fade on the banks of the Taff while I walked and prayed God that he should take a youth from the mine or the field, as he once took Elijah. I prayed that he should not take anyone from Cambridge, lest that feed our pride, nor someone from Oxford lest that subserve our intellectual ambitions. That was the burden of my heart for four years. . . . And now, He who sows and He who reaps can rejoice and say 'Not to us O Lord, not to us, but to Thy name be the glory, for the sake of Thy mercy and Thy truth'.[58]

The Revd. Dr. J. G. Moelwyn Hughes (1866–1944) was the minister of Tabernacle, the Calvinistic Methodist church in Cardigan.[59] He was a quarryman's son from Tanygrisiau in Merioneth. After receiving his education at one of his denomination's theological colleges and taking a pastorate he prepared himself for the entrance examination (in German) of the University of Leipzig. He was successful and took a master's degree and a doctorate of that university. Some years before the revival began it was rumoured both that these degrees had not in fact been conferred upon him and that they were of little value. Eventually he found it necessary to publish two letters in his own defence, the one from the Vice-Principal of the University of Leipzig to the effect that the degrees had been awarded him, and the other from Professor T. Witton Davies of the University College of North Wales, Bangor (who also held a doctorate of the University of Leipzig) to the same effect.[60]

In late January 1905 Seth Joshua conducted a meeting in the

[56] T. M. Rees, op. cit., p. 117.
[57] *The Torch*, vol. 1, No. 1 (New Series), Jan. 1905; 'Notes from my Diary', by Seth Joshua. The pages are not numbered.
[58] *Y Goleuad* (Caernarvon), 20 Jan. 1905.
[59] For Dr. Moelwyn Hughes, see the *Western Mail* (Cardiff), 26 June 1944; *Y Goleuad* (Caernarvon), 5 July 1944 and 12 July 1944.
[60] *Y Goleuad* (Caernarvon), 2 Dec. 1904.

Tabernacle during which he asked all those who were certain of their salvation to stand up. Few did, and the Revd. Dr. Hughes intervened saying that 'he was representing the feelings of many, among the deacons of the church' that 'many of them were believers, but they could not be certain of salvation'.[61] There ensued a dispute between them about 'the doctrine of assurance' which was not brought to an end until one of the deacons intervened.

This instance apart there was no open clash between revivalists and ministers in South Cardiganshire. Elsewhere in Wales there was a more definite opposition between the many ministers and prominent laymen who had not been awakened and those who had been inspired by the revival. The antipathetic feeling aroused persisted for many years.[62]

As has been seen the normal forms of worship were set aside in revival meetings, so to a degree were denominational distinctions. This was commonly the case among the nonconformist denominations even to the point of members of neighbouring Calvinistic Methodist and Congregational churches holding joint revival meetings in each chapel alternately. This happened at Capel Ffynnon and Capel Crugiau, their joint meetings being attended by members of a third chapel, Clos y Graig.[63] At Llandysul the Unitarians joined the other churches in revival meetings, and a number of ministers wished that 'the boundary walls of denominationalism would soon vanish'.[64]

The Bishop of St. David's issued a pastoral letter in which he gave the revival a cautious welcome:

> Our attitude towards it should be an attitude of sympathy, watchfulness, and prayer . . . Sympathy with the spiritual object of the present revival will be in proportion to our sense of our own spiritual need. . . . Underneath all our unhappy divisions there is latent a fundamental spiritual unity among all Christian believers.
>
> It is not without significance that this is primarily a movement of the laity, and especially of young people.[65]

Evans Roberts's meetings at Blaenannerch were attended by the incumbent of the parish (Aber-porth), and he described the meeting that has already been mentioned (p. 235 above) in the Church's

[61] *The Welsh Gazette* (Aberystwyth), 2 Feb. 1905.
[62] W. Nantlais Williams, op. cit., pp. 86–7 and 90–1.
[63] *The Welsh Gazette* (Aberystwyth), 30 Mar. 1905.
[64] Ibid., 1 Dec. 1904.
[65] Quoted in Awstin, *The Religious Revival in Wales*, No. 2 (Cardiff, 1904).

magazine *Yr Haul*.[66] He attended revival meetings in the Calvinistic Methodist chapel at Aber-porth and conducted similar meetings nightly in the parish church. The Church joined in the revival meetings at New Quay, Newcastle Emlyn, Cilcennin, Pencader, and Henllan, while the incumbent of the parish of Llangynllo conducted revival meetings in a church hall at Maes-llyn.[67] At Abermeurig interdenominational revival meetings were held in chapel and church, with the incumbent conducting the meetings in the church.[68] In North Pembrokeshire similar meetings were held at Cilgerran while in the parish of Meline the rector instituted revival services.[69] At Nevern the members of the Calvinistic Methodist chapel sent a deputation to the church asking the church people to join them in revival meetings, and this they did.[70] During the revival too, annual interdenomination open-air prayer meetings were established, as had been done in 1859.

One of the striking features of the revival is the way in which the roles of different members of a congregation were disturbed or reversed while the revival lasted. Just as the sermon was displaced by the revival address so the minister was displaced by the lay evangelist. Those who remembered the revival of 1859 contrasted the prominent place that the sermon had in that and earlier revivals with its relative unimportance in 1904 and 1905 and pointed out that in Welsh nonconformity the sermon was the primary 'means of grace'. They reminded people that the command to *preach* the Gospel was to be in force to the end of time.[71]

The displacement of the minister by the lay revivalist was the case in a simple literal sense; while the evangelist, man or woman, ascended to the pulpit, the minister remained in the deacons' pew. Neither he nor the deacons had control of the meetings, and his part was limited to testing the meeting for converts at the evangelist's behest. In the presence of such a prominent evangelist as Evan Roberts, the minister, even if himself a revivalist, took no part in the meeting unless requested to do so by the evangelist. This was so during the meeting at Blaenannerch which was described on p. 235 above. It was also the case even if the evangelist sat mute

[66] *Hr Haul*, vol. vii (1905), pp. 161–4.
[67] *Cardigan and Tivy-Side Advertiser*, 24 Mar. 1905.
[68] T. Francis, op. cit., p. 326.
[69] *Cardigan and Tivy-Side Advertiser*, 24 Feb. 1905.
[70] Ibid., 16 Dec. 1904.
[71] *Y Geninen*, vol. 13 (1905), p. 98.

in the pulpit virtually throughout the meeting as Evan Roberts did on occasion, for the normal procedures were set aside at the revival meetings. Ministers themselves commented that the official leadership was ignored.[72] At the chapel where Evan Roberts was a member (Moriah, Loughor) the minister, the Revd. Daniel Jones, resigned as a protest against the way in which the revival was proceeding. As his congregation ignored his wishes concerning the way in which meetings should be conducted he considered that his office had been rendered redundant.[73]

The revival brought the young to prominence, and into disagreement with those of the older members who had not been awakened. The young complained of the lack of response among their elders.[74] The older people maintained that a church should be led by those with long experience. The President of the Welsh Congregational Union in 1904, who was the minister of a church in the Teifi valley, pronounced the view that 'To guard the peace of the church, its young people should be content to be followers until they are called upon to be leaders'.[75] Like the young, women too came into prominence locally and nationally, as has been seen. And the revival built a bridge across the denominational divide that normally separated people.

It temporarily established a new and different division within the chapels. It has been said above that in normal times there was a tendency to accommodate relationships within a chapel to those that obtained in daily life according to the manner in which the society was organized to work the land, and which gave standing to farmers vis-à-vis cottagers, men vis-à-vis women, and the elder vis-à-vis the younger people. But during the revival there was an association between those who had been awakened, ministers and laymen, men and women, young and old, regardless of their normal standing and regardless of denominational boundaries, vis-à-vis those who had not been awakened. People became prominent not because they were the official or elected leaders, but because they had been aroused. In the outcome the tendency to accommodate relationships within the chapels to those of everyday life was checked. Relationships within the chapels developed or deepened a measure of independence of relationships without the chapels, and a place was

[72] Y Goleuad (Caernarvon), 28 Apr. 1905.
[73] The Welsh Gazette (Aberystwyth), 19 Jan. 1905.
[74] See, for example, Y Goleuad (Caernarvon), 5 June 1905.
[75] Y Tyst (Merthyr Tudful), 20 Dec. 1905.

kept for those who might otherwise become in Niebuhr's term, 'the Disinherited'.

In this connection it should be noted that there were many revivals in Wales from the eighteenth century to 1904. Some were nation-wide while others remained local. They all had the common features of a heightened consciousness of sin, concern for one's salvation, un-usually sustained and passionate prayer, rejoicing (*gorfoleddu*), and in the later revivals prolonged singing. It is said that Twrgwyn (Rhydlewis) experienced seven revivals between 1792 and 1832.[76] Some of them, as in 1812, affected a large area of Wales, and South Cardiganshire experienced further revivals in 1859, the 1880s, and 1904. The Revd. Edward Parry noted fourteen revivals in various parts of Wales between 1790 and 1892, and the Revd. G. M. Roberts noted sixteen between 1785 and 1904.[77]

Many of the features that were part of the background of the revival of 1904 were specific to the period, the advent of the 'New Theology' which emphasized the immanence of God as against His transcendence, fear of the consequences of Balfour's Education Act of 1902 (against which Joseph Jenkins campaigned), the working of the reformed system of local government, the extended state educa-tional provision, the coming of new forms of leisure-time entertain-ment, but the revival of 1904–5 was also one of a long series of revivals which for generations affected life in the communities that are discussed here.[78]

[76] J. Hughes, *Methodistiaeth Cymru*, vol. ii (Wrexham, 1851), pp. 18–35.

[77] E. Parry, *Llawlyfr a Hanes y Diwygiadau Crefyddol yng Nghymru* (Corwen, 1898); G. M. Roberts, 'Y Cyffroadau Mawr', *Y Goleuad* (Caernarvon), 22 Oct. 1952 and 24 June 1953.

[78] C. R. Williams, 'The Welsh Religious Revival, 1904–5', *The British Journal of Sociology*, vol. iii (1952), pp. 242–59.

X

THE COMMUNITY AND THE
NATION AT LARGE

IT has already been noted that this was an area in which people were interested in one another's affairs, an area in which what happened to people was of general interest, the events of daily life, family affairs, courtships, marriages, deaths. People played guessing games such as 'Guess who's getting married? Guess who's moving house? Guess what house is for sale? That's a puzzle for you'. Till late in the nineteenth century the local weekly newspaper was intended primarily for the gentry, containing such items as 'Metropolitan Gossip', 'Court and Society', 'Human Sacrifice in India', 'Life in Fiji', 'A French view of Lord Brougham', 'In the Maori Country', and 'Antiquarian Discovery in the Crimea': towards the end of the nineteenth and the beginning of the twentieth centuries it came to cater more for the other people of the area, and as its circulation increased among its chief attractions were the 'Births, Marriages, and Deaths' column, and advertisements of the sales of houses and farms, which interested not only those looking for property but gave to others an indication of which households were likely to move, what changes might take place.

It has been noted too that in respect of the manner in which the society was ordered to meet the needs of working the land, people were necessarily interdependent, as they were in other ways besides. This in itself put a limit on extremes of behaviour for to lose other people's co-operation could lead to practical difficulties where work was done by people rather than machines. One of the signs that a farmer was a bad neighbour was that fewer and fewer people were to be seen at his hay harvests. Other people's opinions counted.

Under these circumstances people were particularly sensitive to ridicule, and feared it. Local wits poked fun at people who acted contrarywise to what was the common practice or at people who made fools of themselves as by paying an excessive price for an animal at a fair, making excessive claims for one's own achievements, being on bad terms with one's neighbours, being unduly

I

inquisitive, or habitually using an unusual turn of phrase. They ridiculed too the pettiness of many of the issues about which people were on occasion at loggerheads. Miscreants were lampooned by local poets, and so were people to whom anything unusual had happened under questionable circumstances, as when a farmer overturned his cart on the way home from market and it was suspected that he was the worse for drink. Such lampooning verses were composed to be sung to popular tunes, and were sung and hummed for some time after the event, occasionally for several years. One man purchased a Christmas goose and set off for his home carrying the goose in a sack. He called at a tavern on the way only to find on reaching home that what he had in his sack was not a goose but a cat. Shortly afterwards a string of verses, with chorus, sung to a popular tune, and poking fun at the unfortunate man, was to be heard in the locality. It was sung one Christmas after another for many years. In 1960 I was given by two informants independently of one another, a song of fifteen verses of six lines each, which had been composed in 1857 when a number of nonconformists in one locality had become communicants of the Church of England, under, it was alleged, unfair pressure. This song referred sometimes indirectly, sometimes by name, and in distinctly unflattering terms to the individuals who had thus changed their allegiance. I was requested not to publish the song as the allusions could still be easily understood, more than a hundred years later, both by the descendants of the people ridiculed in the song and by others as well.

Poets poked fun at those who had innocently suffered some happening that could be turned to amusing account, they lampooned those who had transgressed against local opinion. They were on occasion paid to compose verses ridiculing someone who had suffered mishap, and sometimes paid by the sufferers for not composing. One of the most popular features at concerts were topicals. These were composed by a local poet, they alluded to events that had befallen individuals known to the audience, and were sung by a local singer. Certain poet-singer teams were very well known, and could readily amuse an audience. Indeed, some of those who sang the topicals were gifted entertainers. Reference has been made to popular interest in the local newspaper's 'Births, Marriages, and Deaths' column. It contained another very popular column headed 'Week by Week', which consisted of short accounts of untoward events that befell local people. Anyone who suffered in such a way was likely to be threatened with the 'Week by Week'. The newspaper also published lampooning songs on occasion. When people had become tired of one man's endless complaints about the quality of the water supply at a local market town, a song was published which sympathized with the complainant about the poor quality of his water. Needless to say, this kind of thing was only effective where people did constitute a community, were interested in one another's affairs, and where other people's opinions did count.

This sort of ridicule not only exposed the pretensions of some and gave pause to others, but contained an element of debunking, of 'chopping down to size' of people who laid claim to some supposed superiority. The widespread use of nicknames held a similar suggestion. It has sometimes been suggested that this was simply due to the relative paucity of Welsh surnames.[1] But that nicknames were also used in a comparable way in some rural communities in England renders this doubtful. Flora Thompson described how in areas on the border of Northamptonshire and Oxfordshire the men were known by nicknames but their wives as 'Mistress'.[2] Similarly in South Cardiganshire it was generally men who were nicknamed, while women were far less commonly known by nicknames unless they were widows or spinsters. Despite differences of standing and prestige, there was a felt sense in which all were equally members of the same community, that all were in the same boat as it were. While a farmer might be known as Jenkins Rhosdeg and a cottager as Will Penlon, the inequality disappeared when they were referred to by their nicknames while the wives shared their husbands' standings. The point may be made clearer by a recent example. When Princess Margaret and Mr. Anthony Armstrong Jones (as he then was), who was a professional photographer, were married, the story spread like wild fire in South Wales that should they come to live in Wales the Princess would be known as 'Maggie Photos'. The two appellations suggest very different social identities.

It is when people are interdependent that dissensions and quarrels are most serious, for the dissensions and quarrels then affect the work or activity in respect of which the people concerned are interdependent, and it also affects the other people concerned with that work or activity. Where everyone was someone's kinsman or

---

1 D. Parry-Jones, *Welsh Country Upbringing* (London, 1948), pp. 51–4.
2 F. Thompson, *Lark Rise to Candleford* (Oxford, 1939), p. 99.

neighbour or fellow member of a farm work group, or of an inter-farm co-operative group, and where other people's opinions counted, the danger was that differences of opinion might be interpreted in personal terms and genuine disagreements might be seen as personal opposition between those who disagreed. When a regular chapel goer was absent for several Sundays and not because of illness people started speculating with whom he might have quarrelled. On any occasion such as a wedding or a funeral or a farm sale, it was of as much interest to learn who was absent as to know who was present. When people differed about local matters of public policy others wondered what lay behind their differences and considered the alignment of the supporters of either standpoint to see how the issue could be interpreted. Local elections were interpreted as personal contests, and were preceded by speculation as to who would 'come out against' the person seeking re-election. The basic assumption was that issues were to be interpreted in personal terms.

Under these circumstances the maintenance of good relations, or as good relations as possible, was an end in itself and of greater importance than tackling issues likely to prove divisive. Decisions about which people disagreed might be delayed or shelved. Thus when John Crowther (see p. 201 above) and his associates met opposition on offering to provide a wooden building in which to house the library which they had established, and which was at that time housed in the unsuitable premises of an otherwise disused one-time British Society school, Crowther had to state, 'I must be allowed to say that in the judgement of many the opposition was completely unworthy, but rather than create factions and split the district, we chose to let the proposal lapse . . .'.[3]

Attempts were also made to make important decisions unanimous. When a nonconformist church 'called' a minister a double vote was taken, the first to make the actual decision, and the second by pre-arrangement unanimous, to make the formal decision. At the minister's installation it was stressed that the 'call' was unanimous, and by implication that he was not being called to a church that was internally divided and where he could expect the difficulty of conciliating factions who were opposed to each other on personal grounds. In discussing a Norwegian island community Barnes pointed out how such a community might be forced to take decisions that it would have preferred to shelve, when those decisions

[3] *Cardigan and Tivy-Side Advertiser*, 23 Jan. 1920.

were demanded by an external authority (such as the central government) over which the community had no immediate control.[4] But local communities also absolved themselves from the responsibility of raising unpopular issues by insisting that such issues were not of their own choice but forced upon them by an external authority. Thus the Calvinistic Methodist deacon who had to present an issue about which members of the congregation were sharply divided could proceed by saying, 'It isn't we who wish to raise this matter, it is the presbytery that must know'.

When to be a good neighbour involved withholding from actions liable to upset relations between people the divisive issues might be shelved, as in the example given above concerning John Crowther. Or they might be left to be tackled by the person who can best be described as something of an odd man out, the man who in virtue of his social position was sufficiently independent that he could scarcely be accused of being a bad neighbour, or the man who had the temperamental independence to render him immune to such accusations, the man to whom other people's opinions counted for little, or who weighed his neighbour's opinions and was prepared to discount them. Here it may be added that when a community solved its difficulties by leaving unpopular actions to the 'odd man out' there was a price to pay. This was that 'odd men out' were in a position to manipulate affairs to their own advantage. Some of them did so, others did not. The late years of the nineteenth and the early years of the twentieth century saw many changes in the structure of the society and in the society's institutions, many of them disruptive of existing institutions. It was individuals placed as has here been indicated, or of such a temperament as has been indicated, that initiated many of the changes locally.

Above a description has been given of some features of the relationships between people who were closely and necessarily associated with one another in a community wherein the range of daily travel and contacts was limited by the transport facilities available. The character of the relationships can be seen to be consistent with the conditions of living in close and continuous association within the locality, rather than as impressed upon the community from without.

During the nineteenth century there was a steady if irregular

⁴ J. A. Barnes, 'Class and Committees in a Norwegian Island Parish', *Human Relations*, vol. vii, No. 1 (1954), pp. 39–58.

emigration of population from the area, an emigration which from the 1840s onwards was larger than the natural increase, and which came to affect local people's relationships with the nation at large.

TABLE 7

Troedyraur parish: decennial rate at which population decreased[5]

Decennial period	Actual decrease	Rate of population decrease
1831–41	1	0%
1841–51	43	4%
1851–61	46	4½%
1861–71	124	14%
1871–81	14	2%
1881–91	1	0%
1891–1901	63	8%
1901–11	14	2%
1911–21	23	3%
1921–31	129	18%

During the 1850s and 1860s the pent-up excess of population found a mass outlet for the first time as the South Wales coalfield was opened up. This led during the 1870s to the decline of that very important institution, the *neithior* (see p. 131 above), to the decline of the hiring fairs, and to the importation of 'industrial schoolboys' to serve as farm workers. In the last decade of the century the agricultural depression was at its worst, while the parish lost somewhat more than one in six of its population during the 1920s, with the result that in the early 1930s it became impossible for the first time to find tenants for the poorer farms.

Up to a point it was depopulation that allowed the society to remain as it was for as long as it did, for it is most unlikely that the community could have absorbed the whole of its natural increase without the relations between people changing. Only within narrow limits was it possible to cater for the population by adjusting land holding and farm sizes to the population changes, if only because the gentry were the principal landowners, a second conserving factor in the community's life. Comparison of the tithe maps and schedules completed during the years 1836 to 1840 with estates' sales catalogues published immediately before and immediately after the First World War shows that holdings hardly changed at all during

[5] H.M.S.O., *Census of England and Wales, 1831–1931* (inclusive).

these years unless they were in the hands of owner occupiers or very small landowners. But while depopulation relieved the community of the need to accommodate its own natural increase and thereby allowed the society to continue with minimal changes, yet in another way depopulation led to change as it involved integrating the community more completely in the nation at large.

Emigrants opened shops in the mining towns and were supplied with produce from farms occupied by relatives and acquaintances in South Cardiganshire, the goods being transported by horse and cart before the railway was opened to Llandysul in 1864. In addition to this regular traffic, there were families in which the father worked in the collieries, returning to his wife and children during the summer or early autumn when he paid his family's 'work debt' at the corn harvest. This was not a once-for-all migration but continued for some generations and did not cease entirely until the Second World War. At first the people with whom South Cardiganshire communities came into contact in this way were in certain respects not dissimilar to themselves: they were Welsh speaking, sharing a similar 'chapel life', arranged in small-scale if industrial communities, dressed in clothing materials imported from rural factories, eating food transported from the Welsh countryside. But as the mining areas changed so did this contact become a channel for different ideas and practices. The Labour party first found centres that gave it support in South Cardiganshire where family contacts were thus maintained with the East Glamorgan mining valleys for several generations. As South Cardiganshire communities became increasingly integrated into the nation at large so their own social arrangements had increasingly to be accommodated not only to their internal constitution but to their external contacts, and the context of values and ideas within which people thought and acted was modified. The problem here is how to conceive of the relationship between a local community and the nation at large in such a way that it will add to one's understanding of that local community.

The type of society that has been delineated so far may loosely be summed up by describing it as what is sometimes called an 'organic' society. What is meant by this is that one can see the why and wherefore of the social structure in that the ways in which people were related to one another were on a local basis, and that these ways are intelligible when seen in relation to the needs of the land upon which the community lived. This is at best an over-simplification and at worst misleading for no community is self-contained, nor was any community self-contained during the late nineteenth century whatever of any earlier period. Further it may be the nature of a community's relations, cultural, economic, and political, with the nation at large that enable that community to be as it is. It may even be that it is to these 'external relations' that one must look in order to understand certain features of the society rather than solely to local considerations.

In any area some individuals are members not only of local group-ings but of externally based organizations (that is, organizations based outside the locality concerned), whose main strength is, or whose headquarters are, and whose nature, policy, and norms are decided not in the local area but elsewhere. Consequently the tasks that individuals who are members of such organizations are required to perform may be inconsistent with what is expected of them as members of the local community. This can produce problems and tensions not only for the individuals directly concerned but for the fellow members of the local community to which they belong. It may be necessary to modify the arrangements of the local community to cater for such situations, which constitutes a social change, and then the detailed social arrangements will be intelligible not simply in terms of local requirements but of external contacts.

This may be stated abstractly by distinguishing between 'local' social structures, such as farmer–cottager relationships, and 'external' structures, such as the education system, or the police system. It can then be pointed out that the roles to be filled by those who occupy statuses in such external structures may be inconsistent with the roles to be filled by the same individuals in statuses that they occupy in local structures. When discussing a Norwegian island parish where the inhabitants considered themselves to be equali-tarian Barnes described the inconsistencies that existed between the norms of those who considered themselves equals and the roles thrust upon them by a local government system whose hierarchic character was given to it by the central government. Where people considered themselves to be equals they were required to elect some to be representatives while the rest were the represented, the repre-sentatives had to elect officials from among themselves and in doing so to pick and choose between people between whom, on an equali-tarian basis, there was nothing to choose. Barnes proceeded to describe the devices by which an attempt was made to reconcile the norms of the local structures with those of the external structures, such as informal preliminary discussions on a basis of equality that allowed the official council discussions to be merely formal and the re-election of officers *en bloc* to avoid further discussion and further choice.[6] In Central America Redfield noted the inconsistencies between the norms of the educational system as presented in a village school, and those of the village community. While the villagers practised public segregation of the sexes the school was co-educational; while the villagers valued what Redfield called 'productive industry' the village school devoted time to games.[7]

In South Cardiganshire various chapels held their preparatory services before the monthly communion Sunday on a week-day, and their major occasions in the course of the year might occupy the whole day, in preaching festivals, singing festivals, and 'subject' festivals. This was practicable when people were self-employed or employed by people who shared the same norms and obligations as themselves. But as more people came to be members of external organizations their obligations to these organizations were frequently incompatible with their customary obligations to the local community. Postmen had set duties to perform; once the Milk Marketing Board was established milking-times on farms were controlled by the hour at which the Milk Marketing Board lorry arrived, and the lorry drivers were required to be at their duties whatever their other obligations. From about four o'clock in the afternoon farmers were to be seen leaving public gatherings, formal and informal, in time to return home for the milking. I heard one story quoted of the farmer who expressed the incompatibility between what local practice inclined him to do and what milking-time demanded of him in the words 'I'm a slave to that bloody churn'. When external organizations called for Sunday work from someone who had considered himself obliged not to work on Sundays, he was presented with a conflict of loyalties knowing that his neighbours would watch the outcome, and with all the more interest if he was a deacon.

The language of such external organizations is frequently English while many members of the community prefer to speak Welsh. In the case of the Women's Institute the factors of language and

---

[6] J. A. Barnes, op. cit., pp. 39–58.
[7] R. Redfield, *A Village that Chose Progress* (Chicago, 1957), pp. 137–8.

the fact that the Institute's norms are decided outside the local community are both involved, in that the constitution of the Women's Institute requires that the proceedings shall be conducted in the English language. One outcome has been the establishment of branches of a comparable but distinct Welsh-speaking women's movement (*Cymdeithas Merched y Wawr*) in South Cardiganshire during 1967 and 1968. New groupings arise as those who are opposed to being in practice required to speak English recognize an affinity with one another *vis-à-vis* those who are indifferent as to language.

The effects of such influences is not one spectacular change, but cumulative, and often desired. One-time cottagers who felt their inferior position when asking a local employer for permission to attend a meeting or a service not infrequently prefer a more distant connection with an impersonal authority.

It is not the point that such externally based organizations appeared in South Cardiganshire for the first time in the late nineteenth century. The clergy and the nonconformist ministry had in fact long been members of such organizations, if the word be permitted, for instance, while the gentry were as much members of a nation-wide social stratum as they were of the local community. But as changes occurred in the nation at large these changes were channelled into particular communities through the medium of many of the external organizations, a point to be mentioned again below. The increasing integration of South Cardiganshire communities into the nation at large both through the multiplication of, and the changing character of, external organizations, and by the physical links produced by the emigration of people who maintained connections with the people of their native areas, resulted in changes in the community and in the society.

It has been seen that there was an established system by which the land was worked. This system involved farmers and cottagers, with the cottagers forming the work groups attached to each farm. The work groups were of particular importance at the corn harvests. Farms were grouped together to co-operate in some activities, especially those of the hay harvests. On each farm the women cared for the milking, dairy-work, and the farmyard animals, the men of the family for the work of cultivation and the horses. In the organization of men's work sons and servants were equivalent to one another, while the general labouring work was done by hired labourers on those farms where the family could not manage all the work. It has been seen that this constituted not only an organization of farm work, but of relationships between people, of the social system. The introduction of new agricultural methods and of new machinery not only changed the organization of farm work, but also changed those relationships that were connected with the organization of farm work. We shall not be concerned here with all the changes in agriculture that occurred in the late nineteenth and early twentieth centuries, but only with those that affected the relationships between people in virtue of the way in which the land was worked.

The first threshing machine that I have been able to trace was introduced into South Cardiganshire in 1875 or very shortly afterwards. Fixed threshing machines were in use in some parts of Scotland before the end of the eighteenth century, but the first portable threshing machine was not devised until 1803. Such threshing machines did not become common in southern England until the 1840s, these were steam powered but they were drawn from farm to farm not by a steam engine but by horses. It was during the later 1860s that threshing machines powered and drawn by a steam traction engine first appeared. The thresher introduced into South Cardiganshire in or soon after 1875 was a horse-drawn machine and was purchased by a man who had emigrated to Gloucestershire after having served as chief servant on a South Cardiganshire farm.[8] He purchased the machine on his return. (The first threshing machine drawn by a traction engine came to South Cardiganshire about 1905; it was able to serve a larger number of farms than a horse-drawn machine but had no other effect on the organization of farm work than did the horse-drawn machine.) The result of the introduction of the portable threshing machine was not to curtail co-operation between farms but to extend it.

Prior to the coming of the machine each farm threshed its corn independently, either by flail, or by small fixed machines that were generally water powered. It was then necessary to thresh regularly, perhaps daily during the winter, in order to provide feed for the animals and straw for their bedding. (A 260-acre farm employed two men during the winter to do nothing but thresh with the flail.) But once the threshing machine came it was necessary to gather sufficient people together to cater for the machine's demands. Those farms which already co-operated at the hay harvest came to co-operate

[8] Information from the purchaser's son.

with one another when the threshing machine visited them. Co-operation was not immediately and readily extended in this way when the threshing machine first visited one of a group of farms that co-operated at the hay harvest, but over a period of years it became the common practice. The new machines introduced during the period that concerns us did not usually destroy co-operation between farms, but led rather to its elaboration. The potato digger was first patented in 1855, on the principle of rotary spinners that is still in use. It was first introduced into South Cardiganshire shortly before the First World War, and became common during the war, a time of labour scarcity. Before the introduction of the machine potatoes were harvested with the hoe, by the farm's staff and work group. It appears that some farms co-operated for the potato harvest so that the other co-operating farms in a group sent members of their own staffs to help at the harvest of any one of them. But this only became common with the introduction of the machine, for it needed a considerable number of people to keep up with the machine. Farms which co-operated at hay harvests and threshing times came to co-operate at the potato harvest too. It is noticeable that farms never have co-operated to harvest those roots (such as swedes and turnips) which are not harvested by machine.

Co-operation at the hay harvest was more elaborate than on other occasions of co-operative work because it involved mutual help on two different occasions, the mowing, and several days later the carting. Every farm contributed scythesmen to mow at each of the farms in the co-operative group. Generally the order in which farms mowed their hay varied annually according to the condition of the crop at each individual farm each season. But in some co-operative groups farms mowed in the same order each season regardless of variations in the state of the crop. The hay was turned over and prepared for carting by the farm's staff and the women of its work group, but on the day when the hay was to be carted and built into stacks carts were sent from the other co-operating farms to the one which was carting its hay, and the stacks were built by men using pitchforks and rakes but no machines.

When the mowing machine was introduced it became possible for one farmer to upset the mowing arrangements between farms while yet remaining dependent on the other farms' co-operation for carting the hay. Hay mowers were first included at the Royal Agricultural Society's trials in 1857; they were first introduced into

South Cardiganshire about 1890, and became common during the opening years of the twentieth century. Many farmers objected to them for some reasons that have subsequently proved groundless, and for other reasons which were irrational at the time they were stated. Some argued that hay would not grow again where the mowers' iron wheels had run, others simply asserted that the cattle would not eat hay that was machine mowed. Their introduction precipitated a number of fist fights between farmers whose co-operative work at one stage of the harvest was being disturbed, but who would still need to co-operate at the second stage of the harvest. But by the early years of the twentieth century, when mowing machines were more common, some groups of farms turned the machines to co-operative use and whereas they had previously sent scythesmen to mow at other farms they now came to send their machines to mow at the one farm. All of a farm's hay could then be mowed quickly, which made for even drying and a better harvest. The first tedder was devised by Salmon of Woburn in 1830: some were introduced into South Cardiganshire during the 1890s, and during the years 1905–10 some farms procured masts and hooks with which to lift the hay when the stacks were being built. Co-operation for carting the hay remained the normal arrangement until the introduction of quite different machines, such as the baler, after World War II.

But during the years when these machines were introduced a major change in social relationships resulted from the coming of the corn reaper and the self-binder. Bell's reaper was in use in the Carse of Gowrie during the 1830s, and American reaping machines (which reaped the crop but did not bind it into sheaves) were exhibited at the Great Fair of New York in 1851 and at the Great Exhibition in London in the same year. Self-binders that cut the corn and bound it with wire were imported from America in 1873, and a few years later came a machine that cut the corn and bound the sheaves with binder twine. In 1879 McCormick won an award at trials held at Derby, for a machine that bound the sheaves with binder twine, and in the same year the Appleby String Binder appeared. The first reapers to be purchased in South Cardiganshire were in the form of attachments to the mowing machine. This cut the corn and allowed the cut corn to be delivered in suitable quantities for a man or woman to bind it into individual sheaves.[9] Reaping attachments

[9] Verses addressed to 'The Reaping Machine' and dated 1869 are to be found in *Blodau Hefin* (Lampeter, 1883), pp. 30–1. This is the work of the central Cardiganshire poet Dewi Hefin (D. Thomas). It is clear that this was a reaping

were purchased by those farms which owned mowing machines, but within very few years the self-binder was introduced. I have heard claims that one was purchased in the area in 1890, but the first date I consider certain is 1894. Numerous farmers purchased machines at the turn of the century and by 1910 they were common though by no means universal.

Farms' corn-harvest labour needs had customarily been provided by a social device, the work group; with the coming of the self-binder this was replaced by a mechanical device. The knot that bound farmers and cottagers together was severed. Farmers became independent of cottagers in respect of labour needs at the corn harvest, and cottagers became independent of farmers. As shall be mentioned below the corn binder (like other machines) was imported during a decade of marked emigration from the area when the remaining labourers were increasingly able to manage without the perquisites they had previously obtained at the farms where they planted their potato rows. In practice 'setting out potatoes' did not end suddenly once a farm purchased a self-binder. It lasted as a mutually convenient but no longer essential practice, with fewer and fewer people 'setting out potatoes' until, and in some cases even after, the Second World War. But the effect was to sever the heretofore necessary connection between farmer and cottager. Cottagers became increasingly unfamiliar with farm work, and mutual interest in one another's affairs ebbed. Farms no longer commanded work groups and with this the associated status relationship was changed. 'The people of the little houses' in the customary sense came to an end though the phrase may still be heard on occasion when people are discussing the past. People who remember the period are perfectly clear that it was the self-binder rather than any other machine that brought an end to the society that once existed, and when they say so it is to the end of the work groups and the changes in farmer–cottager relationships that they refer.

It has been noted that certain machines affected and extended inter-farm co-operation, while during the same period other machines, notably the reaper and the corn self-binder, made farmer and cottager independent of one another. Other new machines, methods, and marketing organizations affected the work and

machine and not a self-binder. It is less clear whether the author had seen it in use in the area wherein he resided, but as the verses were written as an entry in an *eisteddfod* competition a number of people were presumably familiar with the machine.

organization of a farm's staff. The cream separator (which was devised in Denmark and first shown at the Royal Show of 1879) was brought to South Cardiganshire around 1900, largely as the result of the work in dairy education undertaken by the University College of Wales, Aberystwyth. It rendered redundant the stone bowls in which the milk had previously been stored while the cream rose to the top, and facilitated the work of the dairy. (While men mention the self-binder as the machine that did most to change their work in this period, the women mention the cream separator.) Milking machines did not become practicable till the pulsator principle was adopted, producing an intermittent suction. Such a machine was shown at the Royal Show of 1895. It was in 1913 that it received its first trial at the Royal Show. It was brought into South Cardiganshire in 1917. But the greatest change in dairying was brought about by the change-over from the sale of casked butter to that of bulk sales of milk.

From the time of the agricultural depression at the end of the nineteenth century corn production has declined (apart from increases in cultivation during times of war). Meanwhile the number of dairy cattle kept increased, while the number of other cattle (such as store cattle) fell after reaching a temporary peak in 1901.

TABLE 8

Percentage increase in the number of dairy cattle:
parishes of Troedyraur and Betws Ifan[10]

Period	Actual increase	Percentage increase
1891–1901	21	3
1901–13	39	6
1913–21	54	7½
1921–31	48	5
1931–41	115	14
1941–51	226	24

Little milk was sold from door to door; it was necessary either to sell it in bulk or to make it into butter which was sold to the butter merchants who blended it with imported butter. In the summer of

[10] Figures are given for the joint parishes because the boundary between them was changed during the intercensal period 1931–51, while their joint boundary remained virtually unchanged. Source: data supplied by the Ministry of Agriculture and Fisheries.

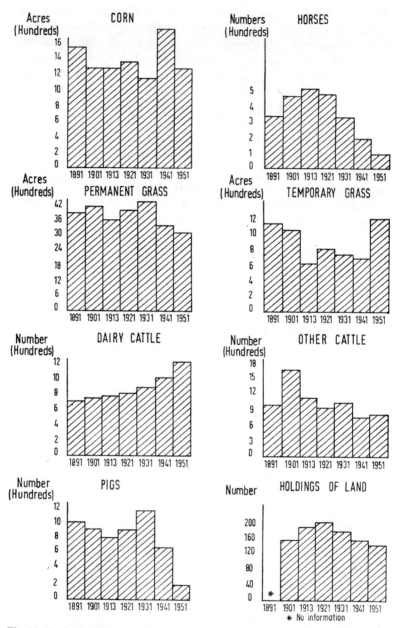

Fig. 15. Cultivation and principal stock; joint parishes of Troedyraur and Betws Ifan.

1927 the butter merchants were paying farmers some 14*d*. per pound (in South Cardiganshire) for their butter.[11] As it requires some two and one-half gallons of milk to make one pound of farm butter, the cash return to the farmer for his milk was 5½*d*. a gallon, in addition to which he had the value of the skim milk and the buttermilk which he retained. Milk was sold in bulk to private dairy companies during the 1920s albeit on a small scale. Depending on the farmers' contracts with the dairy company, the prices that they received for their milk during the summer months of 1932 ranged from 6½*d*. per gallon to 5½*d*. per gallon, and when penalty clauses in their contracts operated, the price might fall to 4*d*. per gallon.[12] (The supply of milk was at its maximum, and the price paid for it at the minimum, in the summer months.) It was with the establishment of the Milk Marketing Board, under the provisions of the Agricultural Marketing Acts of 1931 and 1933, that the sale of milk in bulk became commonplace. The Milk Marketing Board offered a guaranteed market for bulk milk at a guaranteed price considerably higher than the dairy farmers could obtain from private dairy companies. In the summer months of 1934 this price was 1*s*. per gallon.[13] In the following years the number of dairy cattle kept increased at treble the previous rate of increase.

With the sale of bulk milk to the Milk Marketing Board the work of the farm dairy virtually came to an end. Whereas heretofore farmers had received cash only irregularly, from dealers, whenever they had butter or animals to sell, they now received a monthly cheque for the milk they sold to the Milk Marketing Board. Prior to this farmers made use of banks to finance their farming operations (as contrasted with simply having a bank account) only if they were of sufficient standing for banks to advance them money on the strength of promissory notes. Other farmers found it necessary either to ask their better-placed neighbours to stand security for them in order to obtain such loans, or to borrow from them directly. Shopkeepers (who sold seeds and tools as well as groceries) were often local money-lenders. It now became less necessary to obtain goods

[11] *Marketing Milk, Butter, and Eggs in the Counties of Carmarthen, Pembroke, and Cardigan. Being the Report of the Three Counties Marketing Committee, 1931* (Carmarthen, no date), p. 23. This price was the same as the average price obtained for butter at Llandysul in 1877. See J. Llefelys Davies, 'The Diary of a Cardiganshire Farmer', *The Welsh Journal of Agriculture*, vol. x (1934), p. 9.
[12] J. Lewis, *Some Aspects of the History and Development of Dairying in Carmarthenshire*, Appendix J; unpublished M.Sc. thesis in N.L.W. (1948).
[13] H.M.S.O., *Dairy Produce Supplies 1935* (London, 1936), p. 97.

on credit from the shops until the time of fairs at which the dealers paid in cash for the stock they purchased, or later, until the time farmers had stock in condition to sell at the auction marts. And in addition each farmer had to conform with the standards for cattle houses and cleanliness decided upon by the Board's agents in order to be accepted as a supplier to the Board.

As has been stated above in Chapter III, in the past it was the farm women who were closely associated with the milk cattle, while the men of the household were associated with the horses. Men would accept employment as cowmen only if they were unable to find posts as servants, who worked the horses. It was considered an insult for a man to have to milk. Unless the women cleaned the cowstalls the work was done by the second servant or the 'lad', the 'petty servant' who was the lowest in the hierarchy of male farm workers. If there were sons of working age their work was with the horses, not the cattle. Farm maids did much of the work of the farmyard, and it was a mark of superiority that a maid should work in the dairy, with, or instead of, the farmer's wife.

By the time the Milk Marketing Board was established young women were no longer readily available as farm maids. They preferred to emigrate, or to enter what other employment was available. It is a moot point whether the shortage of maids made dairy farmers more ready to become suppliers to the Milk Marketing Board or whether the introduction of bulk milk sales made farm maids redundant. Dairy work came to an end. During the same years tractors were displacing horses. The first tractors came in 1917, in connection with the Agricultural Committees established during the First World War, and the first privately owned tractors soon after the end of the war. The depression of the 1920s and 1930s slowed down the rate at which tractors were purchased, and it was not until the years following the Second World War that they completely replaced horses for farm work. By then the farm maid had virtually disappeared. The only women on farms were farmers' wives, daughters, and sisters, and no daughter would enter another farm but as a wife. As dairy cattle became more important in farm economy, and as the men's prestige animals, horses, became fewer, men undertook the milking and not uncommonly the cowman became the chief servant, or his equivalent if a married man. On the other hand while farm women had not worked with horses farmers' daughters drive tractors. From another standpoint farm organization has changed little, whereas in the past a farm needed a servant to each pair of horses, it now needs a servant to each tractor, if servants are employed at all. But there are so many fewer servants that daughters may drive tractors, and many farmers are more tied to their farms now than they were before the machines we have noted were introduced, for they have fewer servants and dairy cattle (which tie those who care for them to the milking times) are more important than they were.

The common practice of selling milk to the Milk Marketing Board and of receiving payment in the form of a monthly milk cheque led to farmers' independence of local supplies and suppliers of capital and to a greater dependence on the banks. It contributed to loosening the ties of interdependence and obligations between people within a community, and to changing relationships within that community. On farms the change-over to bulk milk sales and the introduction of the tractor eroded the connection between the women and the milk cattle on the one hand, and the men and the horses on the other, which had been one of farm work's most prominent customary features. The whole contributed to changing the attitudes that had been part and parcel of the 'parochial' life that I have tried to present.

An attempt has been made above to indicate the nature of the rural community of the area into which the new machines and methods were introduced. It was a small-scale community in which people were related to each other in customary ways, as farmers and cottagers, as groups of kin and affines, as friends and neighbours. People valued good neighbours, and being a good neighbour involved avoiding words and deeds which were likely to disturb good relations between members of the community. When unpopular issues had to be faced, not infrequently it was left for the 'odd man out' to broach them. The 'odd man out' was described above as the man who in virtue of his social position or temperamental characteristics, was relatively immune to the resentment or adverse criticism his words and actions might provoke. It has been seen too that certain of the machines that were introduced did affect, in one way or another, the customary relations between people.

During the last decade of the nineteenth century emigration from the area was heavier than was usual. Between 1891 and 1901 the population of the parish of Troedyraur fell by 8 per cent, and after 1901 it continued to fall though at a slower rate. The number of farms did not fall, and mechanization apart, their labour needs were little changed, but there were fewer workers available. Small farms avoided haymaking on the same days as neighbouring large farms, for the latter commanded so much of the labour available that there were too few workers to harvest the hay of the other farms without undue difficulty. When haysheds were first built they were not erected on the largest farms in the first place, for those farms could still procure the labourers to thatch their haystacks. Where estates still had their home farms some were among the earliest purchasers of new machinery which they lent to their tenants when the machines were not required at the home farms. But in other cases the home farms could borrow from tenants who had purchased machines, and on such estates no lead was given in the introduction of machinery. Large farms sited near villages and with village labour available were among the last to purchase machinery, apart from the very smallest farms.

But the supply of labour varied not only with the size and location of farms, but with the personality of the farmer. A bad neighbour had more difficulty in obtaining labour than a good neighbour. Many farms were slow to purchase machines because the occupier was personally popular and deferred action that might upset his relations with other people. On the other hand many of those who bought new machines early were people who discounted the opinion of others, and some purchasers were characterized as 'those who could not keep servants'. One of the machines we have noted above was introduced by a man who ran his farm with a staff of 'industrial schoolboys' rather than the normally constituted farm staff. Certain other machines were purchased early on by people who were temperamentally independent, 'odd men out', and innovators in consequence of this quality.

Those who were relatively independent of popular opinion in virtue of their social position, the gentry, contributed most prominently to the agricultural changes of the time chiefly in the fields of agricultural organization and of livestock improvement. During the last decade of the nineteenth century the Tivy Side Stud Society and the Tivy Side Agricultural Society were both run by the gentry. The former provided good stallions to serve mares kept by local farmers. The latter was responsible for the annual agricultural show held at Newcastle Emlyn, where it was intended that the landlords' exhibits would provide standards for the tenantry to aim at. But many squires had by then ceased to maintain home farms, and were not in a position to contribute exhibits to the show.[14] More effective were the efforts of some squires acting individually, as Major Edward Webley-Parry-Pryse of Neuadd Trefawr, and Morgan Richardson, Esq., of Neuadd Wilym. Until the 1890s almost all the cattle, apart from those kept on estates' home farms, were Welsh Blacks, while some squires had introduced Jerseys to their home farms, and were then introducing Shorthorns. In the late nineteenth century Eynon Bowen, Esq., of Troedyraur, secretary to the Tivy Side Agricultural Society, bred Shorthorns at one of his farms until he let it to a tenant during the 1890s, while in the early years of this century Morgan Richardson, Esq. became noted for his Shorthorn herd. But before the First World War began Board of Agriculture officials were taking the initiative in improving the quality of stock by urging farmers to establish livestock improvement societies. By February 1914 the Rhydlewis and District Bull Society was established to provide the services of a bull of approved quality to farmers who held less than 100 acres, the Society being in receipt of a grant from the Board of Agriculture.[15]

During the 1880s most farm horses were Roadsters, sturdily built animals standing some sixteen hands, Colliers, which were stocky animals standing some thirteen hands, and Welsh Cobs. Each farm also kept a donkey which was used to carry water, to take to the shop to fetch whatever was required by the sackload, for other odd jobs, and to turn out to graze with the cattle in the widely held belief that this would prevent abortion in cattle. The first Hackneys and Shires were introduced about 1890. The normal way in which provision was made for mares to be served was for a few farms to keep stallions, which were sent on tour of the other farms in the locality, each in the care of a groom, during May, June, and July of each year. Payment for their services was in cash. The season opened with a parade of stallions at Newcastle Emlyn on the last Friday of April in each year, and with a similar parade at Cardigan on Barley Saturday, the last Saturday of the same month, the traditional end of the sowing season. Thereafter each groom led his charge to different farms, carrying 'stallion cards' showing the stallion's pedigree. A card was given at each farm, where the

[14] Eynon Bowen, Esq., of Troedyraur House, the show's secretary, complained that the gentry did not support the show and failed to exhibit. See *Cardigan and Tivy-Side Advertiser,* 3 Sept. 1897.

[15] *Cardigan and Tivy-Side Advertiser,* 27 Feb. 1914 and 27 Mar. 1914.

servants nailed the card to the stable door to show how many stallions had called there. The groom, who was commonly known as a 'jockey', spent the night at the farm where the stallion had served mares, and went his way the following day, returning to his home infrequently if at all before the season ended. He ate with the servants in the board room, he slept not with the servants in the stable loft but in the house where the best bedroom was frequently prepared for him. Each stallion bore a name, 'Emlyn Squire', 'Prince of Wales', etc., names which were used to nickname farmers whose sexual morality was questionable.

It was in order to improve stock by providing better quality stallions that the gentry supported the Tivy Side Stud Society, and some squires made individual efforts towards the same end. Thus Major Webley-Parry-Pryse hired a good quality stallion each season to tour farms near to his seat at Neuadd Trefawr, the groom returning the stallion to Neuadd Trefawr each week-end. But again, before the outbreak of the First World War committees established under the auspices of the Board of Agriculture were undertaking this work, and a Shire Horse Committee was established in South Cardiganshire in March of 1914.[16] One of the features of the period was the transfer of leadership from the hands of the gentry.

Of the agricultural co-operative societies which have survived in Wales, the first to be registered under the Industrial and Provident Societies Act of 1893 was established at Cardigan and registered as the Vale of Tivy Agricultural Society on 17 February 1902. It pre-dated similar societies at Newcastle Emlyn, Llandysul, Llanarth, and New Quay, by a matter of days. It was largely the work of Augustus Brigstocke, Esq., of Blaenpant, who was supported by others of the gentry of South Cardiganshire. Brigstocke succeeded to the family estate in Cardiganshire and Carmarthenshire in 1901, and came to know of the work of the Agricultural Organization Society which had recently been established to foster agricultural co-operation comparable to that which had been established in Ireland by Sir Horace Plunkett and the Irish Agricultural Organiza-tion Society. Brigstocke's first attempt in this direction came when he formed very small co-operative purchasing societies receiving their supplies from the coastal villages of Cardigan Bay where sailing ships landed cargoes of superphosphates. The project failed and involved Brigstocke in a 'substantial' financial loss, but none

[16] *Cardigan and Tivy-Side Advertiser*, 6 Mar. 1914.

the less he persevered in his efforts.[17] He addressed audiences of farmers, enlisted the help of the Department of Agriculture of the University College of Wales, Aberystwyth, canvassed support, and published a series of articles entitled 'Notes on Agricultural Co-operation' in the local paper, the *Cardigan and Tivy-Side Advertiser*, to explain his purpose and engender interest. A meeting to discuss the possibilities of co-operation was held at the Emlyn Arms Hotel, Newcastle Emlyn (which the gentry had long used as their unofficial assembly rooms) on 20 September 1901, attended by gentry and working farmers. A further meeting was held on 25 October 1901 at the same venue. Major Webley-Parry-Pryse was in the chair, and the meeting was addressed by W. L. Charleton, an Honorary Vice-President of the Agricultural Organization Society. Morgan Richardson, Esq., of Neuadd Wilym (one of the first in the area to advocate co-operation) formally proposed that an agricultural co-operative be established.[18] At further meetings convened by Capt. Jones-Parry of Tyllwyd, and held at Cardigan and Blaenporth (near to Capt. Jones-Parry's seat) it was decided to establish a co-operative at Cardigan. This was duly registered on 17 February 1902, while the Emlyn Agricultural Society Ltd. (of Newcastle Emlyn) was registered a few days later.

Meanwhile Brigstocke persuaded the county councils of Carmarthenshire, Cardiganshire, and Pembrokeshire to send a delegation of landlords and farmers to Ireland to study the development of agricultural co-operatives there. The original intention was that each county should send six delegates but when the Cardigan County Council decided to pay the expenses of four 'Representative Farmers' only, Brigstocke paid the expenses of two others from his own pocket.[19] These were joined by Brigstocke and two specialists and there were in fact nine members each from Cardiganshire and Carmarthenshire, and six from Pembrokeshire when the

[17] H. Jones-Davies, 'The History of the Agricultural Cooperative Movement in Carmarthenshire', *The Carmarthen Antiquary*, vol. 1, Part 2 (1942). In 1902 the author was the chairman of the Carmarthen County Council and a member of the party which visited Ireland as is mentioned below.
[18] *Cambrian News* (Southern Edition), (Aberystwyth), 1 Nov. 1901.
[19] Cardigan County Council: (1) *Minutes of Proceedings at a Quarterly Meeting of the Council held at the Town Hall, Lampeter, on Thursday, the 6th day of February, 1902;* (2) *Minutes of the Proceedings at a Quarterly Meeting of the (Technical Instruction) Committee, held at the Town Hall, Lampeter, on Thursday, the 17th day of April, 1902;* (3) *Minutes of the Proceedings at a Quarterly Meeting of the Council, held at the County Hall, Aberayron, on Thursday, the 1st day of May, 1902.*

delegation visited Ireland in August of 1902.[20] Brigstocke was a member of the committee of the Vale of Tivy Agricultural Society for some years, and the committee's annual meetings were presided over by Major Webley-Parry-Pryse until he left South Cardiganshire on succeeding to the baronetcy and the family estate of Gogerddan. The society succeeded as a co-operative buying organization but attempts at co-operative butter marketing were abandoned at Cardigan, and at Newcastle Emlyn an attempt to market eggs on a co-operative basis failed.

While Brigstocke had been chiefly responsible for the institution of agricultural co-operatives in the area, it was Sir Edward Webley-Parry-Pryse who primarily undertook the work of establishing local branches of the National Farmers Union. This body had been founded in 1908, having developed out of the Lincolnshire Farmers Union, itself founded in 1904. In December 1911 branches were established at Newcastle Emlyn and Cardigan, and the tenants of Sir Edward Webley-Parry-Pryse's estate at Neuadd Trefawr were among their earliest supporters. But Sir Edward died in 1918, while immediately before and immediately after the First World War the gentry were selling their estates. Brigstocke's estate (Blaenpant) of 2,800 acres was offered for sale in May of 1914 except for the mansion and home farm. The property was sold in July of that year excepting one farm of 56 acres and two small holdings. In May 1920 the remaining property was disposed of, including the mansion and home farm.

Certain squires were selling off outlying parts of their estates as early as the 1870s. It was then and during the early 1880s that many erstwhile tenants saddled themselves with debt through purchasing their holdings at the high prices that obtained during a period of agricultural prosperity, only to find themselves obliged to repay their mortgages from incomes earned during years of depression. But it was after the agricultural depression ended and in a period of relative prosperity that the sale of estates began in earnest. The greater part of the Gernos estate was sold in 1907 and in June 1911 outlying parts of Brigstocke's estates in Carmarthenshire were sold.

---

[20] *Report of the Delegates Appointed by the County Councils of Carmarthen, Cardigan and Pembroke, to Study the Working of Agricultural Co-operation in Ireland* (Llanelli, 1902), p. 3. Thousands of copies of this report were printed in English and Welsh, the translation into Welsh being by Sir D. Lleufer Thomas. The copies were distributed through the parish councils and in some cases through the Sunday schools. See also N.L.W., MS. Glyneiddan 93, pp. 151 et seq.

In the following month Sir Marteine Lloyd of Bronwydd sold his small mansion of Cilrhue with adjoining parts of the Bronwydd estate. In 1913 most of the Gwernant estate was sold, and in 1914, as has been mentioned, Brigstocke disposed of the greater part of the Blaenpant estate. In 1918 the remainder of the Gernos estate was sold and in the same year parts of the Llanaeron and Nantgwylan estates were similarly disposed of. The sale of the remainder of Brigstocke's estate at Blaenpant took place in 1920, and the Bowens' remaining properties in Troedyraur were sold in the following year. Whereas the gentry owned fifty properties in the parish of Troedyraur in 1910, at the end of 1921 they owned only eighteen. In the latter year agriculture was still in a prosperous condition, untouched by the depression to come. Influenced by social and political circumstances yet to be mentioned the gentry commonly took advantage of the circumstances in order to sell out. Those who did not do so soon found themselves unable to dispose of their properties profitably, for land prices fell suddenly after the repeal of the Corn Production Act (which had underwritten the price of corn) in 1921.

Once tenants became owners they were in a position, within limits, to solve family problems by reallocating their land. Several holdings which could not have been subdivided by the occupiers while they were tenants were now subdivided in order to establish sons when they came to marry. When freeholders became bankrupt during the depression of the closing years of the nineteenth century it was common practice to split their farms into lots when they were

TABLE 9

Parish of Troedyraur: number of holdings of land[21]

Size of holdings, in acres	Number of holdings		
	1840	1910	1930
1–14	45	66	52
15–29	14	21	24
30–65	16	23	15
66–99	10	9	10
100–150	6	8	10
Over 150	10	6	4
Total number	101	133	115

[21] Sources: Tithe schedules, parish of Troedyraur in N.L.W. (1840); Estates' Sales Catalogues in N.L.W.; personal inquiry.

offered for sale, and this was generally done by detaching a number of cow places from the main area of each farm. Despite the fact that farmers no longer needed a work group as in the past, it was still advantageous for cottagers to gain the 'milk and butter sustenance' that could be provided by a cow place, and the number of these holdings did not fall substantially until the fourth and fifth decades of the twentieth century. Meanwhile there was a continuing place for what have been described above as 'ancillary holdings' in the agricultural system. Whereas the ancillaries of agriculture were once

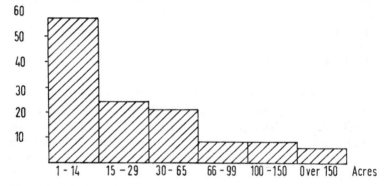

Fig. 16. Parish of Troedyraur, classification of holdings according to size, 1909.

pig butchers, butter and egg merchants, and hauliers who undertook work with horses and carts, they are now agricultural contractors with threshing machines, balers, and cattle lorries available for hire. These ancillary holdings have a place too in relation to the family system, as is explained below.

When estates were sold and farms purchased by owner occupiers, fewer farms were available for rent. And as after the coming of the self-binder an incoming occupier did not have a work group available, he required more capital than in the past if only because he needed to purchase machinery. As a result it became far more difficult for a labourer to obtain a farm and establish himself as a farmer. The gradation in the size of holdings ceased, relatively, to provide a social ladder. On the other hand the relative prosperity of agriculture for some years after the Second World War made it easier for farmers to establish their children on farms, especially as their families are now smaller than was once the case. Farmers are also in a position to use their relative prosperity and the system of land holdings of graded sizes to solve problems of family structure. Sons are established in holdings once occupied by the ancillaries to the agricultural system, enabling each to marry and establish a separate household while employed for wages on the parental farm but borrowing that farm's equipment to work his own holding. Nor is it unknown for the son who remains at home without a regular wage to have all or a number of the dairy cattle in his name, receiving all or a part of the monthly milk cheque as his return, once the father is in receipt of his pension. Farmers, who in the past did not retire from the land, can and do build houses for their retirement enabling a child to succeed to the farm holding. One of the results is that the structural point of tension in the family system, which was detailed in Chapter VI above, is no longer as pressing as it once was.

Just as changes in landownership, in the size of families, in the availability of a greater number of avenues for advancement, and a rise in the general level of prosperity have provided opportunies to solve problems of family structure, so these and other changes have simultaneously affected people's relationships in mutually connected ways.

In the stock-rearing areas of south-west Wales economic recovery from the agricultural depression of the late nineteenth century dated from about 1906 when stock prices started to recover (whereas conditions had improved some years earlier in those regions of Britain given to cereal production). It has already been noted that it was during the same years that the coming of the self-binder led to the dissolution of the traditional type of relationship between farmer and cottager. In practice this meant not only the disappearance of specific groupings of people and of a link which had bound farmer and cottager to one another, but the end of the tension which was associated with the relationship between them.

Difficulties in the relationship between farmer and farmer were eased too with the resolution of the problem that the introduction of new types of farm machinery, in particular the mowing machine, posed for those who constituted co-operative groups of farmers, And once the major sales of estates were completed there was an end to the dissensions that occurred when tenants had to leave when farms were purchased over their heads. If the weft of this particular society was mutual interest and interdependence, the woof was the tension that existed in conditions wherein each person was affected by the consequences of other people's actions. And as the changes noted above rendered people less interdependent, they also released them from the tensions of mutual involvement.

When people had been necessarily involved in one another's affairs though their interests might in many ways have been in conflict, they were given a measure of solidarity in their religious congregations through the workings of the competition that existed between the different chapels and denominations. But when circumstances changed so as to bring their necessary interdependence to an end, one of the props of this competitiveness, namely people's mutual involvement, was undermined, and for this reason among others, congregations became less and less competitive in their relations with one another. Those chapels which had conducted their singing festivals at one of a number of chapels in rotation now came to meet at the same centre annually. The opportunity for one chapel to take pride in providing a better festival than did the others disappeared. Whereas the two chapels (Calvinistic Methodist and Congregational) in Rhydlewis village had assembled on Christmas Day to conduct a subject festival in which each separately recited its own portion of Scripture and was examined on it, they later came to recite the one portion of Scripture together and to be examined on it as one 'school'. Whereas people spoke of such events as having at one time been 'real competition' between congregations deeply interested in one another's affairs, things changed so that 'No chapel today is much concerned about what other chapels do', as I was told. When mutual interdependence declined such institutions as were in part at least dependent on it, could not but be affected.

*   *   *

The leadership that the gentry once exercised passed from their hands during the period that we have been considering, and as a new leadership arose the relationship between the gentry and the rest of the community changed radically. The gentry were, as has been noted, members at the same time of a local community and of a nation-wide stratum of society. Their position in the local community did not rest solely on the fact that they were the landlords, that they were relatively wealthy, that the direct employment they provided on their estates made them the largest employers in their districts, and that they had the power to pick and choose their tenants, considerations which rested largely on a local basis. It rested too on a conjoint basis of law and custom which vested local political power in the social stratum to which they belonged. The Poor Law Act of 1601, as subsequently amended, virtually imposed a rule by local landlords upon the country. This placed local government in the hands of the justices of the peace whose functions were administrative as well as judicial. Assembled in quarter sessions they undertook the work that has latterly been performed by the county councils, exclusive of the additional duties that have been laid upon the county councils since they have been established. The justices were also responsible for the local control of the police force, with power to read the Riot Act. This required that any assembly of twelve or more people to which the statutory proclamation prescribed by the Act was read, disperse within one hour. In connection with their administrative duties in respect of levying the county rate, of public financial expenditure, the administration of the Poor Law and Public Health measures, the maintenance of roads and bridges, they were wholly or partly responsible for the appointment of the necessary officials and other agents. Thus it was to the justices that people went to canvass appointments, as to the county councillors later.[22]

Appointment to the Commission of the Peace was by the Crown, on the nomination of the Lord-Lieutenant, there being a property qualification, namely an income of £100 per annum direct from land, or a prospective income of £300 per annum direct from land. This, with the custom whereby Lord-Lieutenants rarely (in Cardiganshire) nominated those few who did meet the property qualification but were not members of the gentry, effectively limited the magistracy to landowners and their eldest sons, who were thus vested with the power and responsibility of local government. While women were debarred by statute, it was the practice to appoint clergymen to the Commission of the Peace. In 1871 seven of the sixty-three county magistrates for Cardiganshire were clergymen, which

---

[22] This is not an exhaustive account of the magistrates' duties or of the limitations on their power. For details see for instance: M. D. Chalmers, *Local Government* (London, 1883); E. Jenks, *An Outline of English Local Government* (London, 1921); B. and S. Webb, *English Local Government from the Revolution to the Municipal Corporation Act*, vol. i. The Parish and the County (London, 1906).

constituted another link between the Church and the gentry.[23] In the petty sessional division of Troedyraur the Revd. Rhys Jones Lloyd, the uncle of Sir Marteine Lloyd of Bronwydd, acted for many years as the chairman of the Bench, using Stone's *Justices' Manual* as his *vade mecum.*

But the Local Government Act of 1888 established county councils of elected members, and without property qualification. At the same time local control of police forces was allocated in part to the county councils and in part to the magistrates, by means of standing joint committees. And in 1906 under the provisions of the Justices of the Peace Act property qualifications for the magistracy were abolished, and advisory committees established to forward nominations for the Lord-Lieutenants' consideration. This radical change in the position of the gentry of South Cardiganshire was not, however, the outcome of any social change in the local community but of changes in the nation at large and the emergence of an industrial and commercial class whose members were debarred from local government outside the municipal boroughs prior to 1889, regardless of their personal property, unless they owned sufficient landed estate to qualify. That is, one of the major social changes in the local community was the result not of developments as between the gentry and others within the community, but of developments elsewhere which affected the whole stratum of society which the nobility and the gentry constituted. Yet this change in law did not at once and of itself change the society of South Cardiganshire. The new position was that the local government electors were in a position to elect the gentry to the positions to which they had previously been nominated, or to reject them. This choice was no mere matter of logical possibility, for in many areas in Britain at the first county council elections the nobility and the gentry were readily elected, and provided most of the chairmen of the new councils.[24]

The first county council elections were held in January 1889. In Cardiganshire there were eighty-nine candidates for forty-eight seats. Twenty-one of the candidates were members of the gentry: ten were elected and eleven defeated. Whereas in 1888 county government was exclusively in the hands of the gentry, after the election the gentry had ten seats on the county council while the

---

[23] T. Nicholas, *Annals and Antiquities of the Counties and County Families of Wales*, vol. i (London, 1872), pp. 188–9.

[24] F. M. L. Thompson, *English Landed Society in the Nineteenth Century* (London, 1963), pp. 288–9 and p. 325.

others had thirty-eight. The chairman of the new council was a North Cardiganshire coal merchant. When the council elected its first aldermen it chose two who were members of the gentry and twelve who were not.

In South Cardiganshire six squires were among the candidates for fourteen seats.[25] Two were elected, one being unopposed. The election was fought on party lines throughout the county. Only one candidate stood as an Independent and he was defeated. The others stood as Conservatives, Liberals, and Unionists. The squire returned unopposed was W. O. Brigstocke of Blaenpant. Brigstocke had long been a prominent Liberal. He had caught the public eye during the noted parliamentary election of 1868, before the time of his succession to the Blaenpant estate. At that election several of the most influential gentry families in the county favoured the Conservative candidate, Edward Mallet Vaughan, the grandson of the third Earl Lisburne, the family seat being at Crosswood in northern Cardiganshire. Among his supporters were Mrs. Maria Brigstocke, then in occupation of the Blaenpant estate. During the election W. O. Brigstocke walked to the hustings in Cardigan arm-in-arm with one Sam Jones, farmer, to vote for the Liberal candidate, Evan Mathew Richards, who was successful.[26] Sam Jones was a tenant of the Blaenpant estate and according to his own statement he voted for the Liberal candidate (before the introduction of the secret ballot) though the estate's agent had called on him by order of Mrs. Brigstocke to ask him to vote for the Conservative candidate.[27] Mrs. Maria Brigstocke was succeeded at Blaenpant by her nephew, W. O. Brigstocke, as has been mentioned.[28] He in turn was succeeded there by his son Augustus Brigstocke, a Liberal like himself, who as has also been mentioned founded the agricultural co-operative movement in the area.

In the Troedyraur district the Liberals held their political meetings in the Congregational chapel at Rhydlewis. Farmers who would not let their children absent themselves from school during the hay harvest allowed them to do so to attend these meetings. Between 1895 and 1901 the Troedyraur branch of the Cardiganshire

---

[25] By 'South Cardiganshire' is meant that part of the county south and west of a line from Llandysul to New Quay.
[26] *Cardigan and Tivy-Side Advertiser*, 6 Feb. 1920.
[27] *Royal Commission on Land in Wales and Monmouthshire; Minutes of Evidence*, vol. iii (London, 1895), p. 457.
[28] N.L.W., MS. Glyneiddan 92, p. 62 contains a note on W. O. Brigstocke.

Conservative Association had thirty-one officials, committee members, and delegates.[29] Twenty-one of these were churchmen, so that churchmen, though a small proportion of the adult population (see Chapter VIII), constituted some 70 per cent of the local Conservative branch's office-holders.

In the Troedyraur electoral district which consisted of the parishes of Troedyraur, Llangynllo, and Betws Ifan, there were two candidates for one seat in the first county council election. The Conservative candidate was Sir Marteine Lloyd of Bronwydd. Sir Marteine's name had been proposed (whether or not with his consent I do not know) as the Liberal candidate at a meeting held to choose that party's candidate, but this proposal found no seconder.[30] The candidate then adopted by the Liberals was one John Powell, a freeholder, a member of a relatively small kin group and a deacon at the Calvinistic Methodist chapel of Twrgwyn. The chapel had been conspicuously supported for many years both by John Powell's kin and by his affines, though these were again relatively few in number.

The popularity of the Bronwydd family in general, and of Sir Marteine Lloyd in particular, has been noted above (in Chapter I). Though the family ceased to reside at Bronwydd after Sir Marteine's death in 1933 even during the late 1950s some individuals who had recently settled in the area made themselves *personae non grata* by speaking slightingly of the family and their influence in the district. But it was John Powell who was elected and Sir Marteine Lloyd defeated. In the electoral district of Blaen-porth Major-General Alexander Jenkins of Penrallt was defeated; at Llandysul Charles Lloyd, M.A., of Waunifor was defeated. So was C. E. Longcroft, Esq., at New Quay and Llanllwchaearn.

The reality of the transfer of power soon became obvious. The first standing joint committee to control the police force was established during the 'tithe war', when crowds assembled to prevent distraint sales at farms occupied by people who refused to pay tithes. One particularly violent encounter between the crowd and the police occurred at Brynhoffnant in March of 1889 when some forty people were hurt.[31] A force of eighty-four policemen was then brought into the district including members of the Glamorgan Constabulary and

[29] N.L.W. MS. 15857B.

[30] *Cambrian News* (Aberystwyth), 16 Nov. 1888.

[31] H. Evans, *The Cardiganshire Constabulary: Rules, Orders and Guide to Constables* (Aberystwyth, 1897), p. 105.

there was talk of bringing in a military unit to help maintain the peace. Afterwards there came 'the unexpected and unexplained dismissal' of the Chief Constable, Major C. Basset Lewis. When it came to appointing his successor it transpired that twelve members of the standing joint committee supported the payment of tithes while the other twelve (including the chairman) did not. A police superintendent of the Cardiganshire Constabulary was proposed for the post. He was a churchman. Twelve members voted for his appointment and twelve voted against; the chairman exercised his casting vote against the appointment. The same superintendent was proposed a second time with the identical result, and then a third time with the identical result. This was followed by the appointment of an officer of the Carmarthenshire force to be the Chief Constable of the Cardiganshire Constabulary. He was a Calvinistic Methodist.

Meanwhile the basis of appointments to the Commission of the Peace was unchanged and the magistrates were still of the ranks of the gentry as in the past. They expressed dissatisfaction with the way in which the new Chief Constable was exercising his duties, holding that he was using insufficient force to control the crowds which assembled to prevent the execution of tithe sales under distraint orders. But the Chief Constable enjoyed the confidence of the common people. Eventually the magistrates inquired of the Home Secretary as to what powers they retained to issue instructions to the Chief Constable. The reply was that they had couched their inquiry in terms so abstract that no helpful answer could be given and in practice the Chief Constable retained the powers of discretion that he had previously enjoyed.[32] Under the changed circumstances obtaining after the establishment of the first standing joint committee the magistrates' opposition to the Chief Constable was seen to be impotent.

It needed the intervention of the Home Secretary at a later stage to bring the riots to an end. This only came when the police gave more elaborate assistance to the tithe bailiff after the Home Secretary had threatened to see that the financial grant to the county force was withheld unless effective action was taken.

The first elections for the newly constituted parish councils were held in 1894. They were taken very seriously for they were considered to constitute some measure of Home Rule. In Troedyraur parish there were eighteen candidates for twelve seats. The candidates

[32] N.L.W., MS. 15321D, p 5.

included Eynon Bowen, Esq., of Troedyraur House, a member
of a very well-known family in Wales, and the secretary to the Vale
of Tivy Agricultural Society. Among the other candidates was
James Griffiths, the rector's coachman and a Congregationalist who
was released early from his duties to attend his church's regular
prayer meetings wherein he is remembered giving out hymns, two
lines at a time, in the manner that was common before hymn-books
came into use. James Griffiths was elected, Eynon Bowen was not.

It would be entirely mistaken to assume that the gentry candidates
were rejected on personal grounds. They remained as personally
popular after the elections as they had been before. What was rejected was their claim to leadership. This is echoed rather bitterly
in the words of a South Cardiganshire squire, H. M. Vaughan:

> Of the squires that survive today only a few sit on the local Benches,
> the majority of these Benches is nowadays usually composed of
> *ex-officio* members (Rural District chairmen, local Mayors, etc.) and of
> successful traders and profiteers, with a small sprinkling of farmers. . . .
> Modern Democracy has not proved such an unmitigated success that
> our descendants will regard its complete triumph, in the changing
> atmosphere of the end of the nineteenth and beginning of the twentieth
> centuries, as an indisputable blessing, which it would be sacrilege to
> question or even to criticize. . . . All over the world the truth, the real
> truth, is only now beginning to appear above the clouds of democratic
> fiction concerning the French Revolution, the American Civil War and
> other social, economic and political conflicts of the last century and
> more.[33]

That the rejection of the gentry leadership was not inevitable is
clear, if only because there were other areas in which it did not
happen. It was the outcome of the teaching of another leadership,
connected with the nonconformist denominations whose leading
figures had already assumed the cultural leadership of the community. Local nonconformist leaders partook in the prestige of nonconformist leaders elsewhere, for they were all members of a Welsh
nation-wide cultural community. When the gentry lost the mantle
of leadership it was donned by the leading figures of 'the people of
the chapel' and those connected with them. In the outcome this
leadership was comparatively short lived for the community has
been integrated even more thoroughly into the nation at large and
since the Second World War what was a simple emigration of
population has changed into an exchange of population which has

[33] H. M. Vaughan, pp. 3–5.

brought in strangers who do not share in the communal values and attitudes that supported the social life of the past.

*   *   *

The greatest sense of loss experienced by those who lived during the period that I have considered is the loss of a community in which everyone knew everyone else, a community wherein the details of daily life as viewed from a local perspective were matters of common interest and of substance. As the perspective changed with the community's increasing integration into the nation at large what were once seen as matters of importance appear more trivial, and this is felt to be a loss. Whereas the events of everyday life were invested with interest by 'parochial' attitudes and feelings, those same events, which constitute much of the substance of daily life, and now less significant, of less interest, mundane. Yet people are not in general sentimental about the past, but critical. They are aware of the deficiencies of the life that they knew as well as its virtues. They remember that interest in one another's affairs could deteriorate into no interest but in one another's affairs. They remember in particular the physical hardships, the frequent poverty, and the scourge of tuberculosis. They would not have that time back if they could, but they are proud to have lived in it. They see it as a sort of heroic age in which people were associated in and triumphed over adversity and hardship. Their attitude is reminiscent of the words of Tomas Ó Crohan, the Blasket Islander, that 'the like of us will never be again'.[34]

[34] T. Ó. Crohan, *The Islandman* (London, 1934), p. 323.

NOTES TO PLATES

I. PLATE I, ABER-PORTH BEACH *circa* 1875–80

Six ships each of some forty tons burthen are to be seen, most or all of them discharging their cargoes into the awaiting carts. Some fifteen carts are to be seen, and more than twenty horses. A mast and lifting gear were later installed on the cliff top to unload the ships directly, as the horses could not be spared during the hay and corn harvests.

To the left of the sailing ships one herring boat (about thirty feet long) and two ordinary rowing-boats are to be seen. Immediately above the mast of the ship in the right foreground it is possible to distinguish a lime-kiln.

II. PLATE III, MOELON FARMHOUSE *circa* 1900

In the early nineteenth century Moelon was occupied by a freeholder by the name of David Davies who was a descendant of a daughter of the Bowens of Troedyraur House (N.L.W., MS. 15648B). The Davieses were bilingual, churchmen, and in virtue of their family connections they constituted a link between the gentry and the mass of the population. But the experiences of the years during and following the Napoleonic Wars made it difficult for them to sustain this position. Mr. Davies wrote thus to his son, Mr. Daniel Owen Davies:

Moelon March 18th 1816.

As to country news I have not one good news to give you the Agricultural product is quite a Drug, every article the farmer can spare is of little value, —————— farmers cannot pay their rents, in consequence of the high prices of the lands, which rose with the high prices of late years, the landed Proprietors in general all very cruel to their tenants, they —————— give them up to the attorneys, and the attorneys send their Bailiffs with Distress warrants to Distrain upon their poor tenants, —————— Great numbers of tenantry have already given up their Farms against next Michelmas. —————— It is impossible for any man to assist a friend in need, for there is no money to be had, upon any sort of securities; they have of late sent great many debtors to Cardigan gaol. (N.L.W., MS. 15638A.)

A son of the house who became vicar of Llan-non and later of Cenarth was born at Moelon in 1828 but by 1834 Moelon farm had been acquired by Lloyd Williams, Esq., of Gwernant and a tenant installed.

In or about 1856 the tenancy passed to one John Owen, a carrier engaged in taking farm produce by horse and cart to the industrial areas of south-east Wales. He was the son of Owen Owen, a native of Llan-narth (some twelve miles distant) who had come to the district on being granted the tenancy of a twenty-eight-acre holding owned by a Llan-

PLATE I

Aber-porth beach *circa* 1875–80

PLATE II

Cottages at Tre-saith

PLATE III

Moelon farmhouse *circa* 1900

PLATE IV

CARMARTHEN, NOV. 6, 1847.

HAVING lately entered the MATRIMONIAL STATE, we are encouraged by our Friends to make a BIDDING on the occasion on TUESDAY, the 23rd day of NOVEMBER instant, at the Coopers' Arms, Lammas-street; when and where the favour of your good and agreeable company is humbly solicited, and whatever donation you may be pleased to confer on us then will be thankfully received, warmly acknowledged, and cheerfully repaid whenever called for on a similar occasion,

By your most obedient servants,

JACOB DAVIES,

HARRIET DAVIES,

(LATE JAMES.)

₊ THE Young Woman's Father and Mother, (David James, Stone-mason, and his Wife,) and Brothers and Sister, (David, William, Frederick, Charles, and Sarah James,) and also her Brother-in-law and Sister, (George Vaughan, Mariner, and his Wife,) desire that all Gifts of the above nature, due to them, be returned to the Young Woman on the above day, and will be thankful, with her Brother and Sister-in-law, (John and Anne James,) for all favors conferred on her.

JONES AND EVANS, PRINTERS, CARMARTHEN.

Bidding letter (English)
By courtesy of the National Library of Wales

PLATE V

Sir Aberteifi,
Mehefin 2, 1862.

GAN ein bod yn bwriadu ymuno mewn Cyflwr Priod-
asol, dydd Iau, y 12fed o'r mis presenol, bwiadw n
yn mhellach wneud NEITHIOR, yr un dydd, yn Nhŷ
Mam y Ferch Ieuanc, a elwir Blaencwm, yn mhlwyf
Llanfair-or-llwyn. Cerir eich presenoldeb ar y pryd,
a pha roddion bynag a weloch yn dda eu cyfranu, a
dderbynir yn ddiolchgar, ac a ad-delir pryd bynag y
gelwir am danynt,

Gan eich ufudd wasanaethyddion,

DAVID LEWIS,
HANNAH DAVIES.

———

Pob pwython dyledus i'r Mab Ieuanc, ynghyd â'i Dad a'i Fam,
(Thomas a Margaret Lewis, Pantteg,) i gael eu dychwelyd iddo
ef, y Mab Ieuanc, y diwrnod uchod, a byddant yn ddiolchgar,
ynghyd â'i Frodyr a'i Chwiorydd, ei Dadcu a'i Famgu (David a
Mary Lewis, Park), am bob rhoddion ychwanegol.

Hefyd, pob pwython dyledus i'r Ferch Ieuanc, ynghyd â'i diweddar Dad (Thomas Davies), a'i Mham (Hannah Davies), i gael
eu dychwelyd iddi hi, y Ferch Ieuanc, y diwrnod uchod, a byddant
yn ddiolchgar, ynghyd â'i Brodyr a'i Chwiorydd, ei Hewythr a'i
Mhodryb (James ac Anne Evans, Ffosyffin), am bob rhoddion
ychwanegol.

———

J. R. Davies, Argraffydd, Castellnewydd.

Bidding letter (Welsh)
By courtesy of the National Library of Wales

narth man. (In later years another tenant of this holding was a native of Llannarth, having been granted the tenancy by the owner who was again resident in Llannarth.) John Owen married the maid at Gwernant home farm (colloquially known as 'farm yard', a term not used except in this sense). She was the daughter of a carrier occupying a cottage in the locality. After first living separately in their respective parents' homes and then in a cottage of their own, John and Mary Owen moved to Moelon. One of their children (Elizabeth Mary, 1878–1953), novelist and writer of children's books, adopted the pseudonym 'Moelona'. She is referred to in the ninth footnote to Chapter VI and again in the eighth footnote to Chapter VII.

This family moved from Moelon during the 1880s and were succeeded there by a farmer who was also the agent to the Gwernant estate. His son purchased the farm when the estate was sold in 1913, since when it has twice been sold and each purchase has seen a change of occupation at the farm.

III. PLATE V, BIDDING LETTER (WELSH)

The bidding referred to in this letter was to be held in the young woman's mother's house in the parish of Llanfair Orllwyn, which is in South Cardiganshire. The letter requested the repayment of the marriage dues owing to the young man, to his father, and to his mother; it also declared that they, along with the young man's brother and his sisters, his grandfather and his grandmother, would be grateful for all extra gifts.

Repayment was also asked for all marriage dues owing to the young woman, to her late father and to her mother, and it was declared that they along with the young woman's brothers and sisters, her uncle and her aunt would be grateful for all additional gifts.

Prompt repayment was promised whenever it was requested.

BIBLIOGRAPHY

ADAMS, D., *Datblygiad yn ei ddylanwad ar foeseg a diwinyddiaeth,* Wrexham, no date.

—— *Yr Hen a'r Newydd mewn Diwinyddiaeth,* Dolgellau, 1906.

ARENSBERG, C. M., *The Irish Countryman,* London, 1937.

—— and KIMBALL, S. T., *Culture and Community,* New York, 1965.

ASHBY, A. W., and EVANS, I. L., *The Agriculture of Wales and Monmouthshire,* Cardiff, 1944.

AULT, W. O., 'By-Laws of Gleaning and the Problem of Harvest', *Economic History Review,* vol. xiv (1961).

AWSTIN, *The Religious Revival in Wales,* Nos. 1–5, Cardiff, 1904, 1905.

BALLINGER, J., *Gleanings from a Printer's File,* Aberystwyth, 1928.

BANTON, M. (ed.), *Anthropological Approaches to the Study of Religion,* London, 1966.

BARNES, J. A., 'Class and Committees in a Norwegian Island Parish', *Human Relations,* vol. vii, No. 1 (1954).

BEATTIE, J., *Other Cultures,* London, 1964.

BLOIS, H., *Le Reveil au pays de Galles,* Toulouse, c. 1906–7.

BOTT, E., 'The Concept of Class as a Reference Group', *Human Relations,* vol. vii, No. 2 (1954).

—— *Family and Social Network; Roles, Norms, and External Relationships, in Ordinary Urban Families,* London, 1957.

BOWEN, G., 'Traddodiad Llenyddol Deau Ceredigion: 1600–1850' (unpublished University of Wales M.A. thesis, 1943, in N.L.W.).

CLAPHAM, J., *An Economic History of Modern Britain,* vol. i (1926), vol. ii (1932), vol. iii (1938), Cambridge.

COSER, L., *The Functions of Social Conflict,* London, 1956.

CURLE, A., 'Kinship in an English Village', *Man,* vol. 52 (1952).

DAVIES, B., *Y Pulpud a'r Seddau,* Dolgellau, 1909.

DAVIES, G., 'The Industrial School-boy on Welsh Farms', *Social Problems in Wales* (D. Lleufer Thomas *et al.*), London, 1913.

DAVIES, J. L., 'The Diary of a Cardiganshire Farmer, 1870–1900', *Welsh Journal of Agriculture,* vol. x (1934).

DAVIES, W., *General View of the Agriculture and Domestic Economy of South Wales,* London, 1815.

DAVIES, W. J., *Hanes Plwyf Llandyssul,* Llandysul, 1896.

DIETZE, C. VON, 'Peasantry', *Encyclopedia of the Social Sciences,* vol. xii, New York, 1934.

DODD, A. H., *Studies in Stuart Wales,* Cardiff, 1952.

EISENSTADT, S. N., *From Generation to Generation: Age Groups and Social Structure,* London, 1956.

ELLIS, R., *Doniau a Daniwyd*, Llandybie, 1956.

—— *Lleisiau Doe a Heddiw*, Llandybie, 1961.

EVANS, D., *Adgofion*, Lampeter, 1904.

EVANS, D. ARTHEN, *Beirdd Glannau Ceri*, Barry, 1906.

EVANS, D. WYNNE, *Yr Ysbryd Glan a Diweddar—Wlaw y Diwygiad*, Chester, 1906.

EVANS, E. E., *Irish Heritage*, Dundalk, 1942.

—— *Irish Folkways*, London, 1957.

EVANS, E. K., *Fy Mhererindod Ysbyrdol*, Liverpool, 1962.

—— and HUWS, W. P., *Cofiant y Parch David Adams*, Liverpool, 1924.

EVANS, G. E., *Ask the Fellows who Cut the Hay*, London, 1956.

EVANS, H., *Cwm Eithin*, Liverpool, 1950.

EVANS, J., *Hanes Methodistiaeth rhan ddeheuol Sir Aberteifi*, Dolgellau, 1904.

EVANS, S. and ROBERTS, G. M., *Cyfrol Goffa Diwygiad 1904-1905*, Caernarvon, 1954.

FIRTH, R. (ed.), *Two Studies of Kinship in London*, London, 1956.

FOSTER, G. M., 'What is Folk Culture?', *American Anthropologist*, vol. lv, No. 2, Part 1 (1953).

FOX, R., 'Prologomena to the Study of British Kinship', *Penguin Survey of the Social Sciences 1965* (J. Gould, ed.), London, 1965.

FRANCIS, T. (ed.), *Y Diwygiad a'r Diwygwyr*, Dolgellau, 1906.

FRANKENBERG, R., *Village on the Border*, London, 1957.

FRAZER, J. G., *Folk Lore in the Old Testament*, London, 1918.

FREEMAN, J. D., 'On the Concept of Kindred', *Journal of the Royal Anthropological Institute*, vol. 91, Part 11 (1961).

FURSAC, J. R. DE, *Un Mouvement mystique contemporain, le reveil religieux du pays de Galles* (1904–5), Paris, 1907.

FUSSELL, G. E., *The Farmer's Tools 1500–1900*, London, 1952.

GARRARD, M. N., *Mrs. Penn-Lewis, A Memoir*, London, 1930.

GOODE, W. J., *Religion among the Primitives*, Glencoe, 1964.

GRIFFITH, R. D., *Hanes Canu Cynulleidfaol Cymru*, Cardiff, 1948.

HALL, A. D., *A Pilgrimage of British Farming, 1910–1912*, London, 1913.

HASBACH, W., *A History of the English Agricultural Labourer*, London, 1908.

HEATH, F. G., *British Rural Life and Labour*, London, 1911.

HOWELLS, E. J., *The Land Utilisation Survey of Britain*, vol. iii, Part 40 (Cardiganshire), London, 1946.

HUGHES, A., *The Diary of a Farmer's Wife 1796–1797*, London, 1964.

HUGHES, J., *Methodistiaeth Cymru*, vol. ii, Wrexham, 1851.

HUMPHREYS, E. M., *Gwŷr Enwog Gynt*, Second Series, Aberystwyth, 1953.

INGLIS, K. S., *Churches and the Working Class in Victorian England*, London, 1963.

JEFFRIES, R., *The Toilers of the Field*, London, 1898.

JENKIN, T. J., 'Y Parc Llafur (Y Cae Yd)', *Gwyddor Gwlad*, No. 4 (1961).

JENKINS, D. (ed.), *Cerddi Cerngoch*, Lampeter, 1904.

JONES, D. PARRY-, *Welsh Country Upbringing*, London, 1948.

JONES, F. W., *Godre'r Berwyn*, Cardiff, no date.

JONES, J. I., *Yr Hen Amser Gynt*, Aberystwyth, 1958.

JONES, R., *Crwth Dyffryn Clettwr*, Carmarthen, 1848.

JONES, R. B., *Rent Heavens*, London, 1931.

JONES, R. L., 'Changes in the Pattern of Cardiganshire Farming 1908–1958', *Journal of the Royal Welsh Agricultural Society*, vol. xxvii (1958).

JONES, R. T., *Hanes Annibynwyr Cymru*, Swansea, 1966.

JONES, W. H., *Review of the Progress of Tivyside Creameries for the Period 1932–1939*, Aberystwyth, 1940.

LANCASTER, L., 'Some Conceptual Problems in the Study of Family and Kin Ties in the British Isles', *British Journal of Sociology*, vol. xii No. 4 (1961).

LASLETT, P., *The World We Have Lost*, London, 1965.

LEWIS, J., 'Some Aspects of the History and Development of Dairying in Carmarthenshire', unpublished University of Wales M.Sc. thesis, 1948, in N.L.W.

LEWIS, J. PENN-, *The Awakening in Wales*, London, 1905.

LORD ERNLE, *English Farming Past and Present*, London, 1912.

MAIR, L., 'How Small Scale Societies Change', *Penguin Survey of the Social Sciences 1965* (J. Gould, ed.), London, 1965.

MARTIN, D., *A Sociology of English Religion*, London, 1967.

MEYRICK, S. R., *The History and Antiquities of the County of Cardigan*, Brecon, 1907.

MORGAN, J. J., *Hanes Dafydd Morgan Ysbyty a Diwygiad 59*, Mold, 1906.

—— *Cofiant Evan Phillips*, Liverpool, 1930.

MORRIS, L., 'Cardigan Weddings', *The Gentleman's Magazine*, vol. lxi, Part 2 (1791).

MORRIS, W. M., *A Glossary of the Dimetian Dialect*, Tonypandy, 1910.

NADEL, S. F., *The Foundations of Social Anthropology*, London, 1953.

—— *The Theory of Social Structure*, London, 1957.

NICHOLAS, T., *Annals and Antiquities of the Counties and County Families of Wales*, vol. i, London, 1872.

NIEBUHR, H. R., *Christ and Culture*, London, 1952.

—— *The Social Sources of Denominationalism*, New York, 1957.

Ó CROHAN, T., *The Islandman*, London, 1934.

ORWIN, C. S., and WHETHAM, E., *History of British Agriculture*, London, 1964.

OWEN, T. M., 'Some Aspects of the Bidding in Cardiganshire', *Ceredigion*, vol. iv, No. 1 (1960).

PARRY, E., *Llawlyfr a Hanes y Diwygiadau Crefyddol yng Nghymru*, Corwen, 1898.

PEATE, I. C., *The Welsh House*, Liverpool, 1944.

PHILLIPS, D. M., *Evan Roberts, the Great Welsh Revivalist and his Work*, London, 1906.

PHILLIPS, E., 'Hunangofiant', *Y Geninen*, vol. 28 (1910); vol. 29 (1911).

PHILLIPS, R., 'Some Aspects of the Agricultural Conditions in Cardiganshire in the 19th Century', *Welsh Journal of Agriculture*, vol. 1 (1925).

POLLOCK, J. C., *The Keswick Story: the Authorized History of the Keswick Convention*, London, 1964.

REDFIELD, R., *The Little Community: Viewpoints for the Study of a Human Whole*, Chicago, 1955.

—— *Peasant Society and Culture*, Chicago, 1956.

—— *A Village that Chose Progress*, Chicago, 1957.

REES, A. D., *Life in a Welsh Countryside*, Cardiff, 1950.

REES, T. and THOMAS, J., *Hanes Eglwysi Annibynol Cymru*, vol. ii, Liverpool, 1872.

REES, T. M., *Seth and Frank Joshua, The Renowned Evangelists*, Wrexham, 1926.

Report of the Committee on the Employment of Children, Young Persons, and Women in Agriculture, First Report, London, 1869; *Third Report*, 1870.

Report of the Delegates Appointed by the County Councils of Carmarthen, Cardigan and Pembroke to Study the Working of Agricultural Co-operation in Ireland, Llanelli, 1902.

Report of the Royal Commission on Labour: The Agricultural Labourer, vol. ii (Wales), London, 1893.

Report of the Special Assistant Poor Law Commissioners on the Employment of Women and Children in Agriculture, London, 1843.

ROBERTS, E., *Pedair Safon o'r Bywyd Ysbrydol*, Leicester, no date.

ROBERTS, G. M., 'Y Cyffroadau Mawr', *Y Goleuad*, 22 Oct. 1952 and 24 July 1953.

ROBERTS, J., *Methodistiaeth Galfinaidd Cymru*, London, 1931.

RODERICK, A. J. (ed.), *Wales through the Ages*, vol. ii, Llandybie, 1965.

ROSSER, C. and HARRIES, C., *The Family and Social Change*, London, 1965.

Royal Commission on Land in Wales and Monmouthshire, Evidence, vols. i and ii (1894); vols. iii and iv (1895); vol. v, *Report and Appendices, Index* (1896), London.

SALAMAN, R. N., *History and Social Influence of the Potato*, Cambridge, 1949.

SAUNDERS, S. M., *Y Diwygiad ym Mhentre Alun*, Wrexham, 1907. A novel of the Revival by Mrs. J. M. Saunders.

SIMPSON, E. S., 'Milk Production in England and Wales: A Study in the Influence of Collective Marketing', *The Geographical Review*, vol. xlix (1959).

SMITH, L. P., *Words and Idioms, Studies in the English Language*, London, 1925.

THOMPSON, F., *Lark Rise to Candleford,* Oxford, 1939.

THOMPSON, F. M. L., *English Landed Society in the Eighteenth Century,* London, 1963.

VAUGHAN, H. M., *The South Wales Squires,* London, 1926.

WEST, J., *Plainville,* New York, 1945.

WILLIAMS, C. R., 'The Welsh Religious Revival, 1904–5', *The British Journal of Sociology,* vol. iii, 1952.

WILLIAMS, D., *The Rebecca Riots,* Cardiff, 1955.

WILLIAMS, W. M., *The Sociology of an English Village; Gosforth,* London, 1956.

WILSON, B. R., *Religion in Secular Society,* London, 1966.

—— *Sects and Society,* London, 1961.

—— (ed.), *Patterns of Sectarianism,* London, 1967.

INDEX

Adams, Revd. David, 199, 217 and n.
Affines, 174.
Agricultural co-operation: establishment of co-operative societies, 266; Vale of Tivy Agricultural Society Ltd., 266; Emlyn Agricultural Society Ltd., 267; delegation to Ireland, 267–8.
Agricultural holdings: interdependence of, 42–8, 61; ancillary holdings, 43, 270; size of, 269–70.
Agricultural machinery: introduction of—threshing machine, 255; potato digger, 256; mowing machine, 256–7; reaper, 257; tedder, 257; self binder, 257–8; cream separator, 259; milking machine, 259; tractor, 262.
Agriculture, changes in, 254–63.
Alltyrodyn, 13.
Antrim, County, 157.
Arensberg, C. M., 112, 118, 121, 136, 146, 151.

Bardd Horeb, see Thomas, Evan.
Barnes, J. A., 248, 252.
Battersby, Canon T. D. Harford-, *see* Harford-Battersby, Canon T. D.
Bidding: 127; procedure at, 130–5; 'near relatives' and the, 172.
Board of Agriculture, 265.
Board room (*rwm ford*), 97.
Bondage, 70 and n.
Bowen (family) of Troedyraur House: James, 22; Thomas, 22; Arthur James, 23, 30, 208; Eynon George, 23, 265 and n., 278; John, 33.
Brigstocke, Augustus, 266–8.
Brigstocke, William Owen, 275 and n.
Brongest, 16.
Bronwydd, 18. *See also* Estates.
Burial, procedure at, 4, 215.
Butter, sale of, 261 and n.

Calves, naming practices, 104.
Capel Dewi creamery, 39.
Cardigan county council: establishment of, 274; elections to, 274–6.
Cattle: store, 39, 77; dairy, 76, 259.
Cereals: sowing season, 14, 46; harvest season, 14, 48–51, 52.
Cerngoch, see Jenkins, John.
Charles, Revd. Thomas, 202.
Chief Constable (of Cardiganshire), appointment of, 277.
Colby, Thomas (Pantyderi), 18–19.
Commission of the Peace, 273.
Competition (for farms), 175–6.
Conservative Association, Troedyraur branch, 275–6.
Co-operative groups, *see* Farm co-operation.
Corn drying, 85.
Corn Production Act, Repeal of, 269.
Cottagers ('people of the little houses'), 12–13.
Courtship: night visiting, 126; 'fetching and drawing', 127–8.
Craftsmen, 92, 103.
Crowther, John N., 200, 201, 248.
Cymanfa bwnc, see Subject festival.
Cymanfa ganu, see Singing festival.
Cymdeithas Merched y Wawr, 254.
Cynnos, 85.

Daily routine (of farm work), 98–100.
Dairy-work, 76, 84, 261, 262.
Daniel Ddu, *see* Evans, Revd. Daniel.
Davies, Revd. David (Castellhywel), 34.
Davies, John Humphreys, 227.
Davies, Revd. J. Rhys, 221.
Davies, Revd. Rhystyd R., 229.
Davies, Robert Joseph (Cwrt-mawr), 227.
Debt butter (*menyn dyled*), 53.
Decisions shelved, 248.
Deilad, 44–5, 54.

chell, 232, 233–4; Cardigan, 232, 234, 238–41; Cilgerran, 232, 242; Gorseinon, 233; Loughor, 233; Llandysul, 241; Nevern, 242; conventions at—Keswick, 219, 221, 222; Llandrindod Wells, 221–2, 223, 224, 227; New Quay, 227; Blaenannerch, 228, 230.

Rhydlewis, 16, 17, 182, 201.

Richardson, Morgan: and stock improvement, 265; and agricultural co-operation, 267.

Ridicule, 245–7.

Roberts, Evan: 228, 229–43 *passim*; at Twrgwyn (Rhydlewis), 229–30; at Blaenannerch, 230, 235–6; at New Quay, 230, 236–7.

Roberts, Revd. G. M., 244.

Saunders, Revd. John M., 227, 228, 239.

Saunders, Mrs. Sarah M., 227, 228.

Seed hay, 87–8.

Settlement pattern, 6.

Shimbli, 85.

Siblings, 105, 172–4.

Siencyn, Nathaniel, 34.

Siencyn, Sion, 34.

Singing festival, 197, 202–3, 204–5.

Smith, Robert P., 220, 221.

Social categories, 12–13.

Social change, 5.

Social field, unitary, 189–90.

Social structure, 5.

Sources: documentary, 2, 36–8; oral, 2–5 *passim*.

Stable loft, 123–4.

Stallion cards, 81.

Status, hierarchy of, 60–1.

Stock, improvement of, 264–6.

Structures: external, 252–4; local, 252–4.

Subject (*pwnc*): 197, 198–9; festival, 197–202.

Succession to holdings: 115–16, 146–54, 175; freeholders, 147–50, 152; tenants, 148–51.

Taliesin, 196.

Teetotalism, 213–14.

Temperance festival, 203.

Temperature, seasonal, 15.

Tenants: number of, 147; succession to holdings, 148, 151, 153–4; and inter-farm movement, 177.

Thickens, Revd. John, 224–8 *passim*.

Third Report of the Committee on the Employment of Children, etc., in Agriculture (1870), 76.

Thomas, Evan (*Bardd Horeb*), 34.

Thomas, Siencyn, 34.

Thompson, F., 70, 247.

Threshing, 101–2, 255–6.

Tithe war, 276–7.

Tory Island, 154, 166.

Tradition: 'great', 33; 'little', 33; literary, 33–5.

Tre'r-ddôl, 196.

Troedyraur House, 22–4. See also Estates.

Troedyraur parish: 8, 16–17; population movement, 8; population, 16–17; land ownership, 19–20, 36, 39; population, birthplaces, 158; places of worship, 180–1; parish church, 184–6; diaconate in nonconformist chapels, 190–1.

Tyler, Gwinnet, 24–5, 31.

Ultimogeniture, 153.

Vaughan, H. M., 18–19, 116, 278.

Webley-Parry-Pryse, Major Sir Edward: 265; and stock improvement, 266; and agricultural co-operation, 267; and the National Farmers Union, 268.

Williams, David Pryse, 235.

Williams, G. Lloyd, 21, 22, 30.

Williams, W. M., 153.

Williams, Revd. William (Pantycelyn), 191.

Wills, 149–50.

Women's Institute, 253–4.

Work debt, 52, 53, 58, 61, 63–7 and 67 n. See also Potato setting.

Work group (*medel*), 54–6, 71, 117–18, 258.

PRINTED IN GREAT BRITAIN
AT THE UNIVERSITY PRESS, OXFORD
BY VIVIAN RIDLER
PRINTER TO THE UNIVERSITY